Two Manuals Combined into One Book

Light Antiarmor Weapons

And

Tow Weapon System

Department of Defense

Pentagon
Publishing

FIELD MANUAL
NO. 3-23.25 (FM 23-25)

HEADQUARTERS
DEPARTMENT OF THE ARMY
WASHINGTON, DC, 30 August 2001

LIGHT ANTIARMOR WEAPONS

CONTENTS

*This publication supersedes FM 23-25, 17 August 1994.

PREFACE

This manual provides technical information, training techniques, and combat techniques of light antiarmor weapons. Intended users include leaders and designated soldiers who will use this information to successfully integrate light antiarmor weapons into combat operations. This manual discusses gunnery training and train-the-trainer.

The tactical positions shown in this manual were drawn to enhance the reader's understanding of related subject material and do not necessarily represent the best employment option for that material.

Unless this publication states otherwise, masculine nouns and pronouns do not refer exclusively to men.

The proponent of this publication is the United States Army Infantry School. Send comments and recommendations on DA Form 2028 (Recommended Changes to Publications and Blank Forms) directly to—

Commandant
US Army Infantry School
ATTN: ATSH-IN-S3
Fort Benning, GA 31905-5595
lusanoh@benning.army.mil

CHAPTER 1
INTRODUCTION

This chapter provides information common to the light antiarmor weapons discussed in this manual. Topics include care and handling, destruction and decontamination procedures, and operating temperatures. Light antiarmor weapons are used against light armored vehicles, field fortifications, or other similar targets. These weapons are issued as rounds of ammunition to individual soldiers in addition to their assigned weapons and the unit's organic antiarmor weapons. Light antiarmor weapons can withstand extreme weather and environmental conditions, including arctic, tropical, and desert. The light antiarmor weapons category includes both light antiarmor and light antitank weapons.

1-1. TYPES OF LIGHT ANTIARMOR WEAPONS

Light antiarmor weapons include the M72-series light antitank weapon (LAW) and the M136 AT4. The M72-series LAW was designed in the early 1960's for use against light tanks of that era (Figure 1-1). The M136 AT4 was designed in the late 1980's for use against the improved armor of light armored vehicles (Figure 1-2).

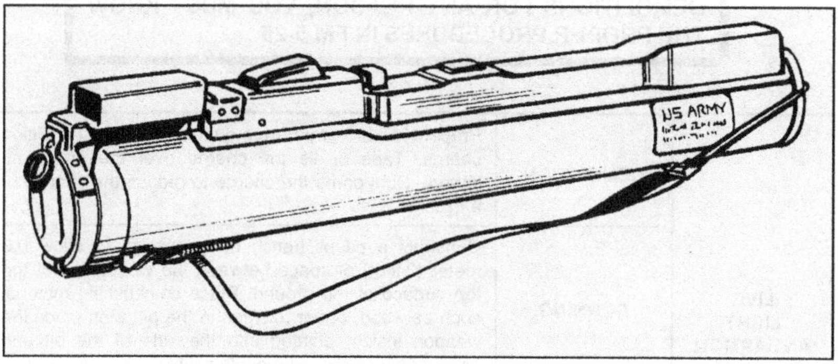

Figure 1-1. M72-series light antitank weapon.

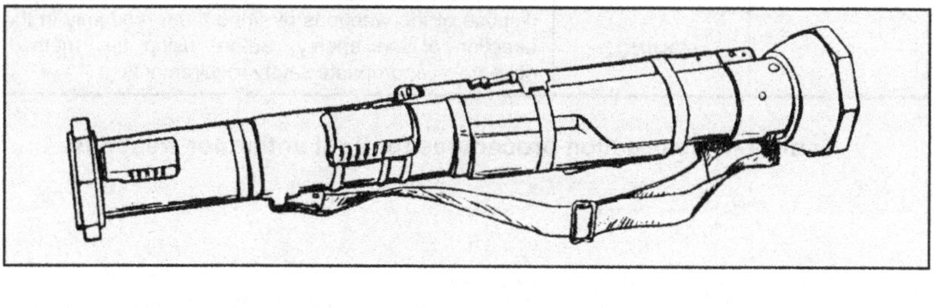

Figure 1-2. M136 AT4 light antiarmor weapon.

1-2. CARE AND HANDLING

Light antiarmor weapons are issued as rounds of ammunition. The only requirement for their care is a visual inspection, outlined in the appropriate chapter for each weapon (Chapter 2 for the LAW and Chapter 3 for the AT4).

1-3. DESTRUCTION PROCEDURES (COMBAT ONLY)

In combat, live and expended light antiarmor weapons are destroyed only to prevent their capture or use by the enemy and, even then, only on order. For such an order to be given, the weapons must be so badly damaged that neither repairs nor cannibalization can restore them to usable condition (FM 5-25). Table 1-1 provides destruction procedures for live and expended light antiarmor weapons; Appendix A discusses safety precautions to follow when destroying them.

DANGER

TO AVOID POSSIBLE INJURY OR DEATH, MOVE TO A SAFE POSITION AND TAKE COVER BEFORE USING ANY DESTRUCTION PROCEDURE. BEFORE USING DEMOLITIONS FOR ANY REASON, YOU MUST KNOW THE PROPER PROCEDURES IN FM 5-25.

LIVE LIGHT ANTIARMOR ROUND	DEMOLITION	Prepare a 113-gram (one-quarter pound) demolition charge. Tape or tie the charge over the propellant charge. Dual prime the charge to reduce the chance of a misfire.
	BURNING	Construct a pit or trench deep enough to allow 0.6 meter (2 feet) of space between the weapons and the top surface of the ground. Place combustible material such as wood, paper, or rags in the pit, then place the weapon inside, pointed into the side of the pit and directed away from all friendly soldiers. Pour diesel fuel or oil over the weapons and the combustible material.
	FIRING	If time does not permit use of the previous methods, dispose of the weapons by firing them randomly in the direction of the enemy. Before using this method, observe all appropriate safety requirements.

Table 1-1. Destruction procedures for light antiarmor weapons.

EXPENDED LIGHT ANTIARMOR ROUND	**MECHANICAL**	Though you must never use mechanical means to destroy live antiarmor weapons or expended M136 AT4s, you may do so to destroy the residue from an expended M72-series LAW. For example, you can drive over it with a tracked vehicle or strike it with a pick, ax, or other object, as long as you make it unusable.
	DEMOLITION	Same as a live round.
	BURNING	Same as a live round.

Table 1-1. Destruction procedures for light antiarmor weapons (continued).

DANGER

1. WHEN USING FIRE TO DESTROY A LIGHT ANTIARMOR WEAPON, THE TIME REQUIRED TO EXPLODE THE WARHEAD IS UNPREDICTABLE. ALSO, IGNITING THE PROPELLANT CAN CAUSE IT TO FIRE THE WARHEAD IN ANY DIRECTION, WHICH COULD IN TURN CAUSE INJURY OR DEATH.

2. OBSERVE THE APPROPRIATE SAFETY PRECAUTIONS WHEN HANDLING DIESEL FUEL. CARELESSNESS COULD CAUSE PAINFUL, EVEN FATAL, BURNS.

3. DO NOT TRY TO USE VEHICLES OR MECHANICAL MEANS TO DESTROY LIVE ANTIARMOR WEAPONS. EITHER METHOD COULD DETONATE THE WARHEAD OR PROPELLANT CHARGE, WHICH COULD CAUSE INJURY OR DEATH.

1-4. DECONTAMINATION PROCEDURES

The soldier can use his M258A1 or DKIE (XM280) individual decontamination packet to remove H-series, G-series, and V-series agents. FM 3-5 provides more information about decontamination procedures for equipment and weapons.

DANGER

NEVER USE DS2 TO DECONTAMINATE ANY LIGHT ANTIARMOR WEAPON. THE DS2 WOULD DISSOLVE THE RUBBER AND PLASTIC SEALS, ALLOWING THE DS2 TO REACH THE PROPELLANT AND PRODUCING AN EXTREMELY HAZARDOUS MIXTURE.

1-5. OPERATING TEMPERATURES

Operating temperatures for the M72-series LAW and M136 AT4 are -40 degrees to 140 degrees Fahrenheit (-40 degrees to 60 degrees Centigrade). Firing light antiarmor weapons in temperatures outside these limits could cause a misfire or produce some other hazard for the soldier (Appendix A).

CHAPTER 2

M72-SERIES LAW, OPERATION AND FUNCTION

This chapter provides information on and technical data for the M72-series light antitank weapon (LAW). It also discusses the characteristics, nomenclature, functioning, and operation of the LAW.

2-1. DESCRIPTION

The M72-series LAW is a lightweight, self-contained, antiarmor weapon consisting of a rocket packed in a launcher (Figure 2-1). It is man-portable, may be fired from either shoulder, and is issued as a round of ammunition. It requires little from the user—only a visual inspection and some operator maintenance. The launcher, which consists of two tubes, one inside the other, serves as a watertight packing container for the rocket and houses a percussion-type firing mechanism that activates the rocket.

a. **Outer Tube**. The trigger housing assembly (which contains the trigger assembly) is on the upper surface of the outer tube. So are the trigger arming handle, front and rear sight assemblies, and the launcher's rear cover.

b. **Inner Tube**. The inner tube telescopes outward toward the rear, guided by a channel assembly that rides in an alignment slot in the outer tube's trigger housing assembly. The channel assembly also houses the firing pin rod assembly, which includes a detent lever assembly. The detent lever assembly moves under the trigger assembly in the outer tube, locking the inner tube in the extended position and cocking the weapon. All this must occur before the weapon can be fired.

c. **Rocket**. The rocket is a percussion-ignited, fin-stabilized, fixed munition. The rocket is attached by the igniter to the inside of the launcher. The rocket consists of a 66-mm HEAT warhead, a point-initiating, base-detonating fuze, and a rocket motor. Six spring-loaded fins are attached to the rear of the rocket motor. These fins are folded forward along the motor when the rocket is in the launcher. When ignited, the propellant in the rocket motor burns completely, producing gasses about 1,400 degrees Fahrenheit (760 degrees Centigrade). The gas pressure pushes the rocket toward the target and exits to the rear of the launcher as the backblast.

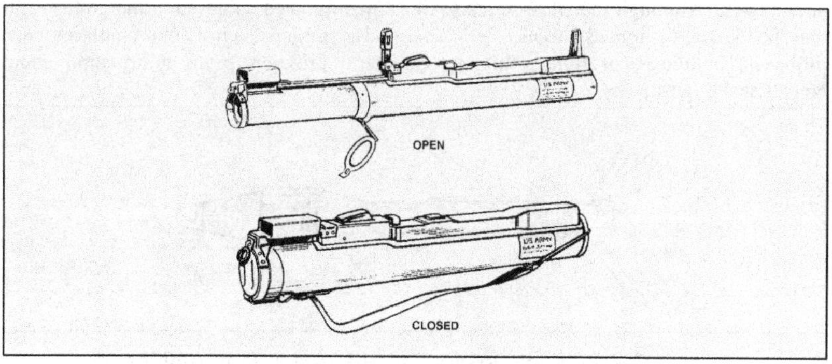

Figure 2-1. The M72A2 or M72A3 LAW.

2-2. TECHNICAL DATA

The following data apply to the M72A2 and M72A3 LAWs:

a. **Launcher**.

Length (Extended)........Less than 1 meter (34.67 inches)
Length (Closed)........0.67 meter (24.8 inches)
Weight (Complete M72A2)........2.3 kg (5.1 pounds)
Weight (Complete M72A3)........2.5 kg (5.5 pounds)
Firing Mechanism........Percussion
Front Sight........Reticle graduated in 25-meter range increments
Rear Sight........Peep sight adjusts automatically to temperature change

b. **Rocket**.

Caliber........66 mm
Length........50.8 cm (20 inches)
Weight........1.8 kg (2.2 pounds)
Muzzle Velocity........144.8 mps (475 fps)
Minimum Range (Combat)........10 meters (33 feet)
Minimum Arming Range........10 meters (33 feet)
Maximum Range........1,000 meters (3,300 feet)
Maximum Effective Ranges:
Stationary Target........200 meters (660 feet)
Moving Target........165 meters (541 feet)
(Beyond these ranges, there is less than a fifty percent chance of hitting the target.)

2-3. AMMUNITION

The M72-series LAW is issued as a round of ammunition. It contains a nonadjustable propelling charge and a rocket. Every M72-series LAW has an integral high-explosive antitank (HEAT) warhead. The warhead is in the rocket's head (or body) section. The fuze and booster are in the rocket's closure section. The propellant, its igniter, and the fin assembly are in the rocket's motor. No inert versions are available (Figure 2-2). Appendix B provides information about appropriate gunnery training devices and ammunition. Although the M72-series LAW is mainly used as an antiarmor weapon, it may be used with limited success against secondary targets such as gun emplacements, pillboxes, buildings, or light vehicles. (Chapter 6 provides more information about combat techniques.)

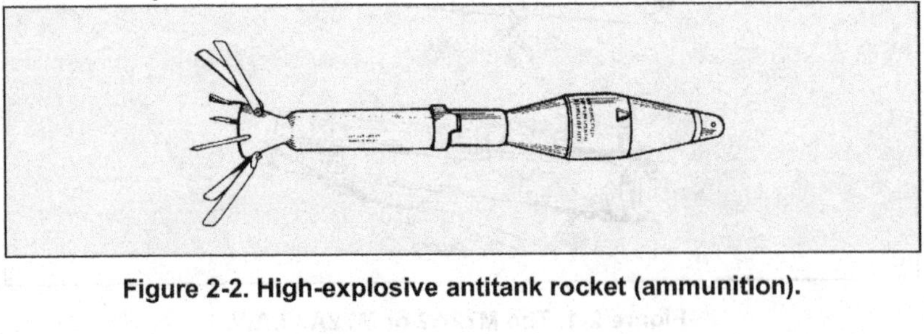

Figure 2-2. High-explosive antitank rocket (ammunition).

a. **Description**. The 66-mm HEAT rocket warhead consists of a tapered, thin-gauge steel body. When it explodes, the force and heat of the explosive focus into a small but powerful gas jet. This directional jet penetrates the target and, if the target is a vehicle, sprays molten metal inside. If the jet hits an engine or ammunition, it may start a fire or cause an explosion. Figure 2-3 shows how the warhead penetrates 300 millimeters of rolled homogeneous steel armor.

(1) ***Impact***. The nose cone crushes; the impact sensor activates the fuze.

(2) ***Ignition***. The ogive crush switch activates the electric detonator. The booster detonates, initiating the main charge.

(3) ***Penetration***. The main charge fires and forces the warhead body liner into a directional gas jet that penetrates armor plate.

(4) ***After Armor Effects (Spalling)***. The projectile fragments and incendiary effects produce blinding light and destroy the target's interior.

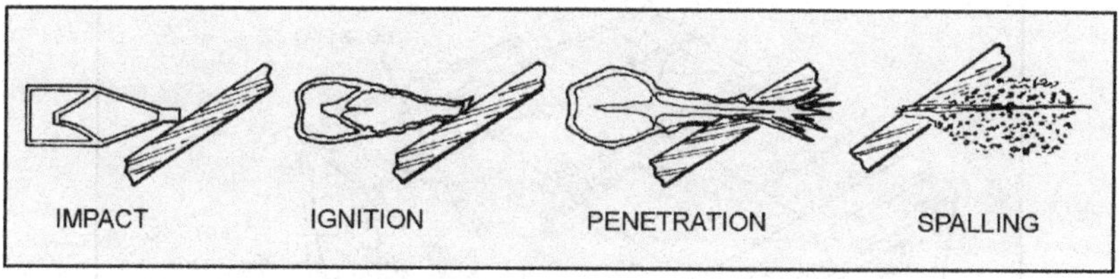

Figure 2-3. Effects of M72-series LAW warhead.

b. **Characteristics**. The head of the round is olive drab stenciled in yellow. The M412 fuze is dropsafe and boresafe. Its minimum arming distance is about 33 feet (10 meters). Six stabilizing fins are attached as part of the motor. As the rocket clears the launcher, springs force open the fins, which stabilize the rocket in flight.

c. **Packaging**. Five complete M72-series LAWs are packaged within a fiberboard inner pack for a total weight of 12.5 kilograms (27 1/2 pounds). Three inner packs are then placed in a wire-bound wooden box, the gross weight of which is 54.5 kilograms (120 pounds) (Figure 2-4, page 2-4).

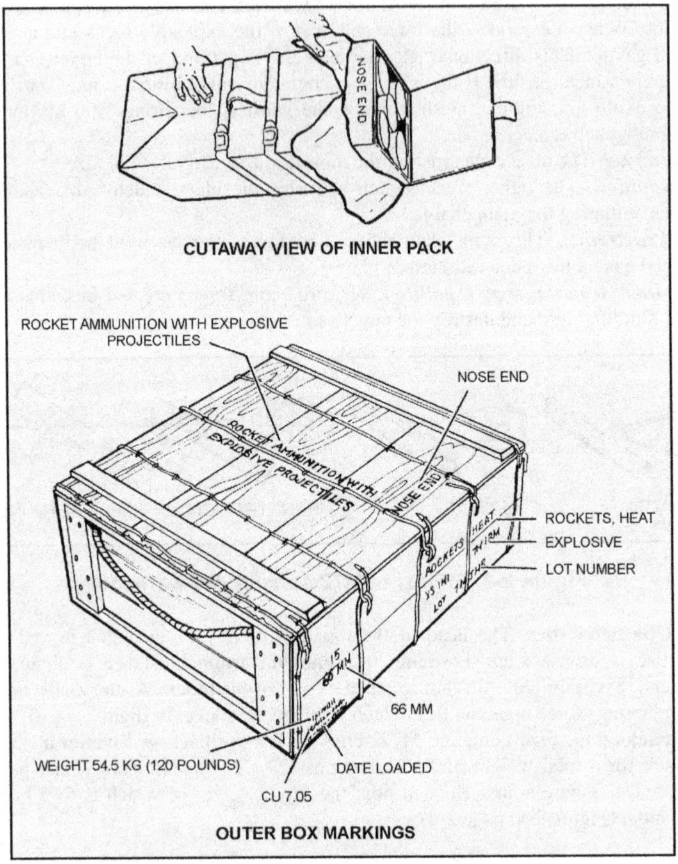

Figure 2-4. Packaging for M72-series LAW.

2-4. INSPECTION

Because the M72-series LAW is issued as a round of ammunition rather than as a weapon, inspection is limited to a visual examination of the sealed unit.

Inspect the launcher's overall condition before preparing the launcher for use.

- Check the body for dents, cracks, or bulges.
- Check the rubber boots covering the trigger bar and barrel detent for tears or punctures.
- Ensure the arming handle is present and on SAFE and that the pull pin is in place.

- Check the data plate for the phrase, W/COUPLER (Figure 2-5).

DANGER

1. IF THE M72A2 LAW DOES NOT STATE "W/COUPLER" ON ITS DATA PLATE, TURN THE WEAPON IN TO THE UNIT AMMUNITION SECTION.

2. THE COUPLER PREVENTS THE INNER AND OUTER TUBES FROM SEPARATING AND POSSIBLY CAUSING PREMATURE DETONATION.

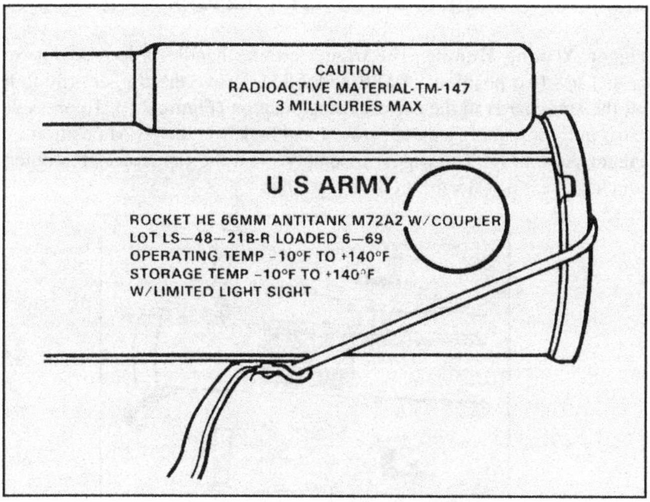

Figure 2-5. Launcher data plate.

2-5. FIRING MECHANISM

The firing mechanism includes the trigger arming handle, the trigger assembly, and the firing pin rod assembly (Figure 2-6).

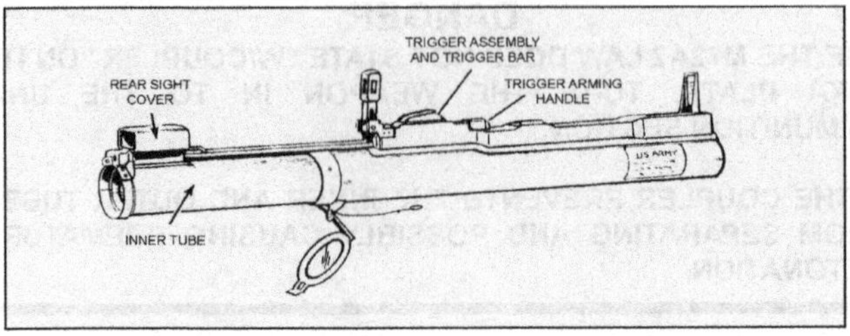

Figure 2-6. Firing mechanism.

a. **Trigger Arming Handle**. The trigger arming handle is located forward of the trigger bar and has two positions: SAFE and ARM. Leave the trigger arming handle on SAFE until the launcher is in the correct firing position (Figure 2-7). To press the trigger, you must first pull the arming handle forward and lock it in the ARM position.

b. **Trigger Assembly**. The trigger assembly is on the top rear of the outer tube. To fire the launcher, press downward on the trigger bar.

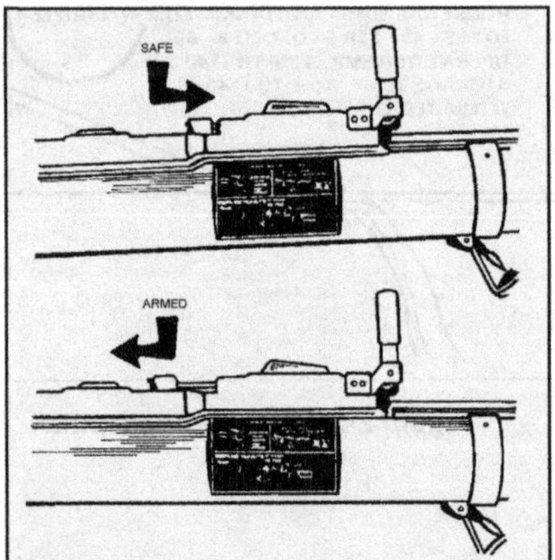

Figure 2-7. Trigger arming handle.

c. **Firing Pin Rod Assembly**. The rear sight cover and the firing pin housing are on the top of the rear of the inner tube. Inside the housing, the primer and the firing pin rod are aligned (Figure 2-8). Pressing the trigger bar releases the tension on the firing pin rod assembly, allowing the firing pin to strike the center of the primer.

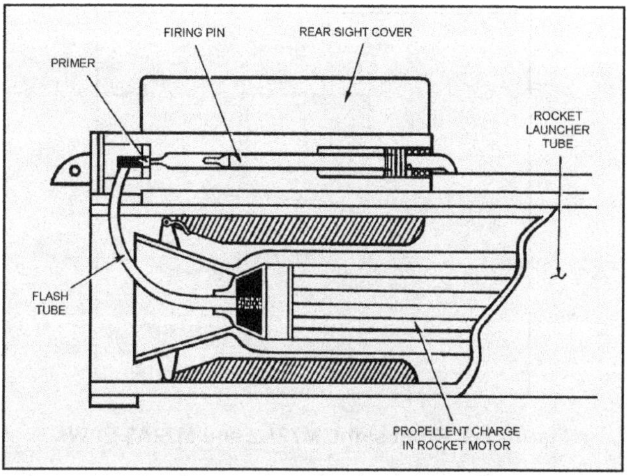

Figure 2-8. Firing pin, primer, and propellant charge.

2-6. SIGHTS

This paragraph discusses the front and rear sights and their proper use.

a. **Front Sight, M72A2 and M72A3 LAWs.** The front sight has a raised vertical range line marked with ranges from 50 to 350 meters in 25-meter increments (Figure 2-9). Two curved stadia lines are etched on the front sights. Do not use the stadia lines on this sight to estimate range, because they are inaccurate. Lead indicators are located on either side of the stadia lines to help you engage moving targets. On the M72A3, use the front sight illuminated range marks at the 100-meter and 150-meter points to help you engage targets in low light.

WARNING

DO NOT TOUCH THE RANGE MARKS; THEY ARE ILLUMINATED WITH PROMETHIUM, WHICH IS MILDLY RADIOACTIVE.

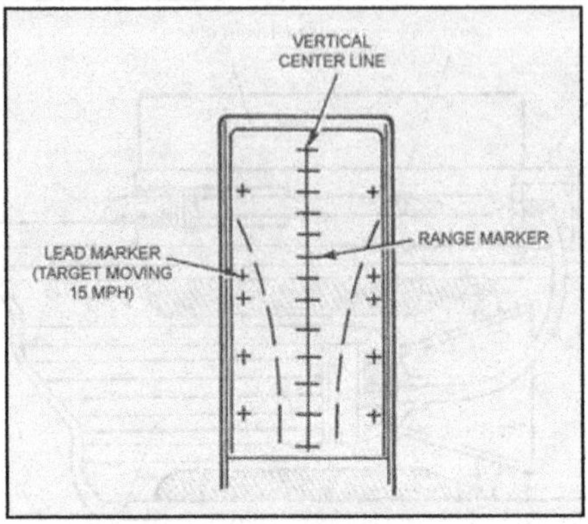

Figure 2-9. Front sight, M72A2 and M72A3 LAWs.

b. **Rear Sight, M72A2 and M72A3 LAWs.** The rear sight consists of a steel bracket with a rubber boot and plastic peep sight. This sight automatically adjusts to changes in temperature (Figure 2-10), which means that its settings are unaffected by temperature.

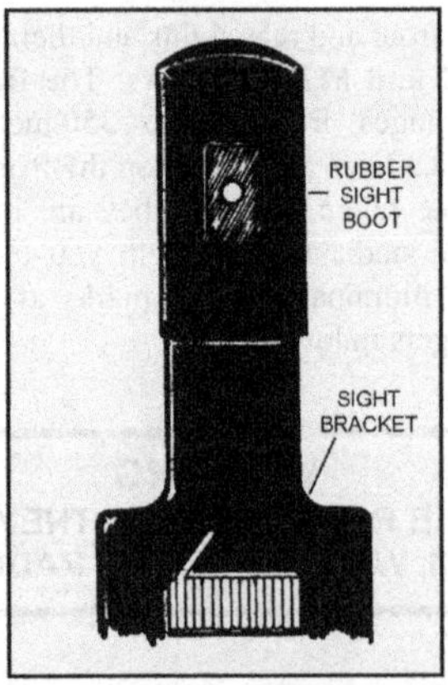

Figure 2-10. Rear sight, M72A2 and M72A3 LAWs.

c. **AN/PVS-4 Nightsight.** This nightsight is issued with various accessories, including a bracket that, when mounted on an M72-series LAW, will allow you to use an AN/PVS-4 on the LAW. However, this works only if DS maintenance has already installed an M72A1 reticle in the AN/PVS-4. Though the reticle was developed for the M72A1 LAW, you can also use it with the other M72-series models. To mount the M72A1 bracket assembly on any M72-series LAWC

(1) Place the bracket assembly on top of the rocket launcher (Figure 2-11) so that the square cutout in the top of the bracket fits over the extension release button.

(2) Swing the lower adapter section up and under the rocket launcher and secure it by turning the locking latch clockwise to fully engage the latch shoulder screw.

(3) Place the sight in the groove on the bracket and align the threaded screw hole in the base of the sight with the lever screw assembly. Tighten the lever screw assembly firmly.

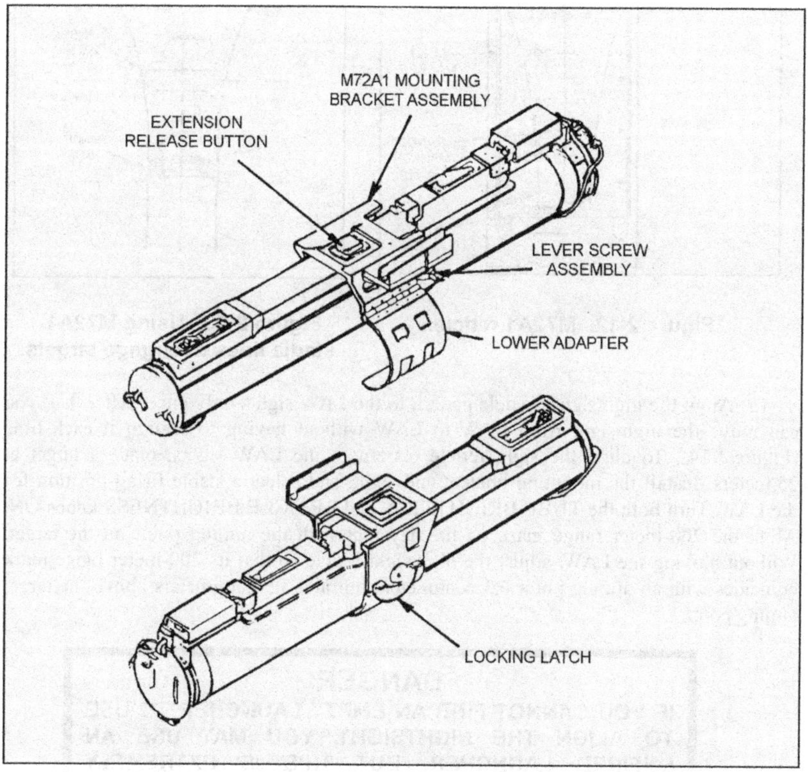

Figure 2-11. Installing M72A1 mounting bracket assembly.

(4) Use the M72A1 reticle to estimate the range to a 20-foot target such as a tank viewed from the side (Figure 2-12). Place the sight so that the rear and front of a flanking target fit between the curved vertical lines (Figure 2-13). When you read the range from the scale, note that the width of the tank is about one-half the tank's length. This means you can estimate the range to a target, whether it is headed straight toward you or straight away from you, by placing its sides between the curved vertical lines and halving the range shown at that point. The bottom of each vertical line in the center of the reticle

corresponds to an additional 25 meters. The length of a horizontal line represents 5 mils, and the spaces between lines represent 5 mils. Use the stadia lines to estimate range *only* when using the AN/PVS-4's M72A1 reticle pattern.

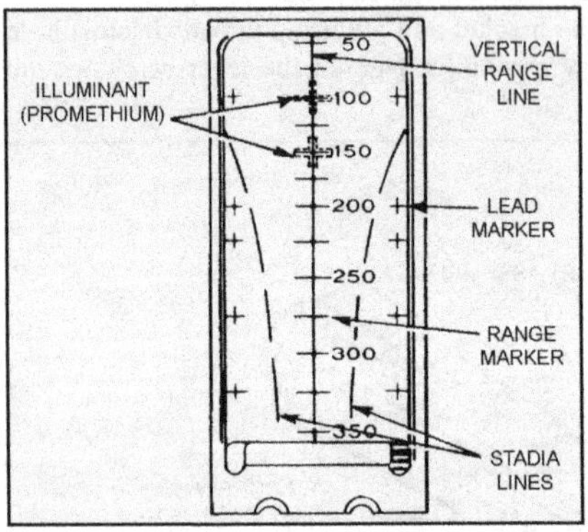

Figure 2-12. M72A1 reticle.

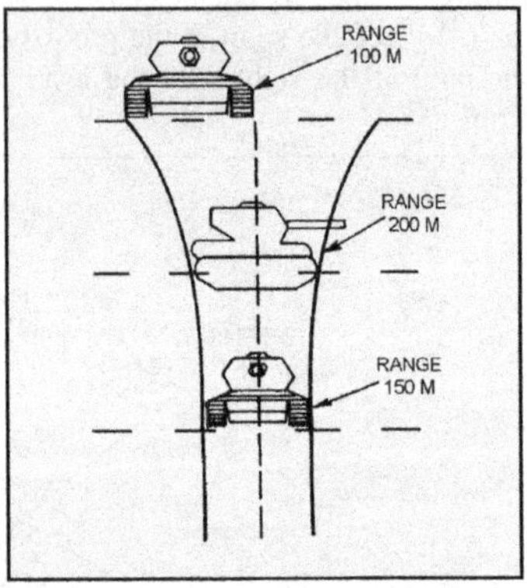

Figure 2-13. Using M72A1 stadia lines to engage targets.

(5) Align the nightsight's reticle pattern to the LAW sights only once. After that, you can move the nightsight from LAW to LAW without having to realign it each time (Figure 2-14). To align the sight reticle pattern to the LAW sights, place a target at 25 meters. Install the mounting bracket and sight, and select a stable firing position for the LAW. Turn both the TUBE BRIGHTNESS and RETICLE BRIGHTNESS knobs ON. Align the 200-meter range mark on the daysight with the aiming point on the target. Without moving the LAW, adjust the nightsight reticle so that its 200-meter range mark coincides with an aiming point 6.3 centimeters left and 3.5 centimeters above the target aiming point.

DANGER

IF YOU CANNOT FIND AN EMPTY LAUNCHER TO USE TO ALIGN THE NIGHTSIGHT, YOU MAY USE AN UNFIRED LAUNCHER, BUT THIS IS EXTREMELY DANGEROUS.

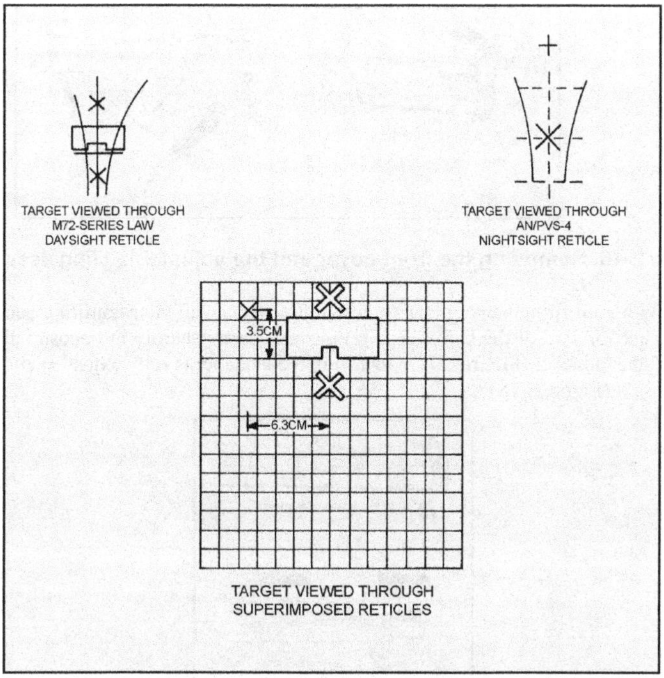

Figure 2-14. M72A1 alignment target for the AN/PVS-4.

2-7. OPERATION AND FUNCTION

Before preparing the launcher for use, the firer inspects its overall condition.

a. To extend the rocket launcher—

(1) Remove the pull pin and rotate the rear cover downward so the front cover and adjustable sling assembly can fall free (Figures 2-15 and 2-16, page 2-11). Do not discard the sling assembly until after you fire the rocket.

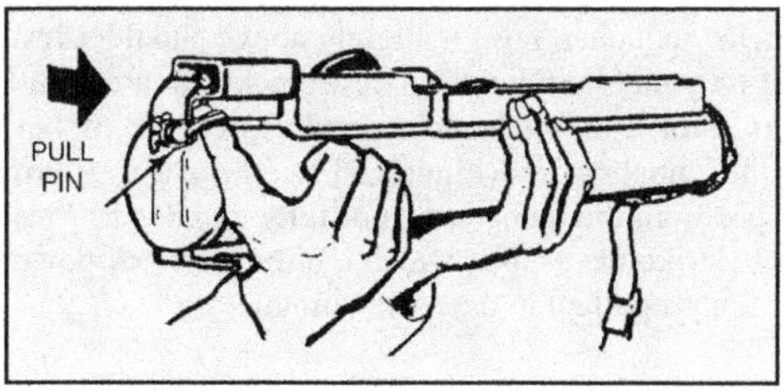

Figure 2-15. Removing the pull pin.

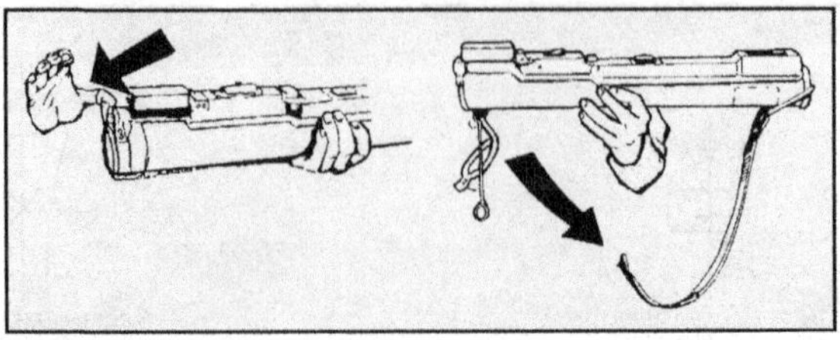

Figure 2-16. Removing the front cover and the adjustable sling assembly.

(2) With your firing hand, grasp the rear sight cover; with your nonfiring hand, grasp the launcher forward of the barrel detent. Pull your hands sharply in opposite directions to extend the launcher (Figure 2-17). To ensure the launcher is fully extended and locked, try to close it (Figure 2-18).

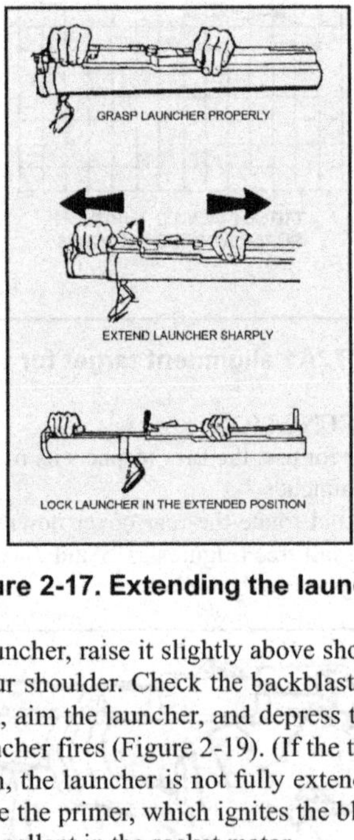

Figure 2-17. Extending the launcher.

b. To fire the rocket launcher, raise it slightly above shoulder level, rotate your body under it, and place it on your shoulder. Check the backblast area, pull the trigger arming handle to the ARM position, aim the launcher, and depress the rubber boot on the trigger bar firmly to ensure the launcher fires (Figure 2-19). (If the trigger arming handle will not remain in the ARM position, the launcher is not fully extended.) Pressing the trigger bar causes the firing pin to strike the primer, which ignites the black powder in the flash tube, which in turn ignites the propellant in the rocket motor.

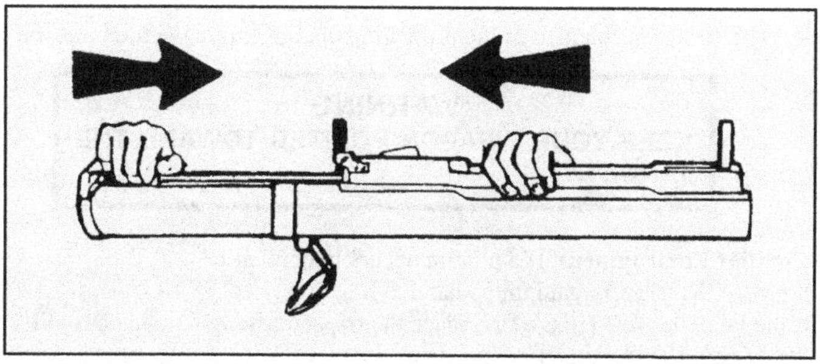

Figure 2-18. Ensuring launcher is locked in the extended position.

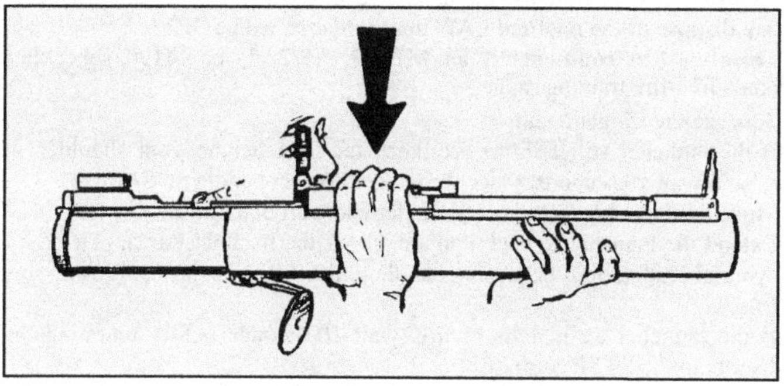

Figure 2-19. Firing the launcher.

DANGER

WHEN OPERATING THE LAW, KEEP IT POINTED DOWNRANGE. ENSURE YOUR WHOLE BODY IS CLEAR OF THE MUZZLE AND REAR OF THE LAUNCHER, AND ENSURE THE BACKBLAST AREA IS CLEAR.

2-8. MISFIRE PROCEDURES

A misfire is a complete failure to fire caused by a procedural or mechanical failure. Which misfire procedures should be used depends on whether the firer is in a combat or training environment.

a. **Causes**. A misfire is usually caused by one of the following factors:

- The launcher may not be fully extended.

- The trigger arming handle may not be armed.
- The firing mechanism or the propelling charge explosive train may be faulty.

> **WARNING**
> **KEEP YOUR WEAPON POINTED TOWARD THE TARGET.**

b. **Combat Environment**. If a misfire occurs in combat—

(1) Squeeze the trigger again immediately.

(2) If the launcher still fails to fire, place the trigger arming handle on SAFE.

(3) Partly collapse the launcher, then extend it to cock it again.

(4) Place it on your shoulder, check the backblast area again, then arm, aim, and fire.

(5) If the LAW still fails to fire, squeeze the trigger again and return the trigger arming handle to SAFE. Collapse the launcher, set it aside, and try another one. As soon as you can, dispose of the misfired LAW in accordance with SOP.

c. **Training Environment**. If an M72A2, M72A3, or M190 subcaliber device misfires on a live-fire training range—

(1) Squeeze the trigger again.

(2) If the launcher still fails to fire, keep the launcher on your shoulder, announce "Misfire," and wait 10 seconds. Place the trigger arming handle on SAFE.

(3) Move the launcher from your shoulder and wait one minute.

(4) Extend the launcher to cock it again, check the backblast area, place the launcher back on your shoulder, pull the arming handle to the ARM position, aim, and squeeze the trigger bar.

(5) If the launcher again fails to fire, wait 10 seconds before returning the trigger arming handle to the SAFE position.

(6) Keep the launcher trained on the target area at least one minute; DO NOT collapse the launcher.

(7) Move the launcher to a safe area and dispose of it IAW unit SOP.

2-9. RESTORATION TO CARRYING CONFIGURATION

If the launcher is prepared to fire, but then is not fired, it should be returned to the carrying configuration by reversing the preparation procedure. After the launcher has been prepared for firing, it is no longer watertight. Therefore, when carrying the launcher, sling it over either shoulder with the muzzle (forward) end down. Only the rocket and rocket motor ignition system are waterproof.

a. Return the trigger arming handle to the SAFE position.

b. Remove the launcher from your shoulder, depress the barrel detent, collapse the launcher tube, and guide the front and rear sights into position.

c. Close the rear cover, replace the cover pull pin, and replace the sling assembly.

> **WARNING**
> **TO PREVENT INJURY, REMOVE YOUR THUMB FROM THE DETENT AFTER COLLAPSING THE LAUNCHER 1/2 TO 1 INCH.**

CHAPTER 3

M136 AT4, OPERATION AND FUNCTION

This chapter provides information and technical data for the M136 AT4 light antiarmor weapon, including its characteristics, nomenclature, and operation. Its function, firing mechanism, and safeties are also discussed.

3-1. DESCRIPTION

The M136 AT4 is a lightweight, self-contained, antiarmor weapon. It consists of a free-flight, fin-stabilized, rocket-type cartridge packed in an expendable, one-piece, fiberglass-wrapped tube (Figure 3-1). The M136 AT4 is man-portable and is fired from the right shoulder only. The launcher is watertight for ease of transportation and storage. Though the M136 AT4 can be employed in limited visibility, the firer must be able to see and identify the target and estimate the range to it. Unlike the M72-series LAW, the M136 AT4 launcher need not be extended before firing.

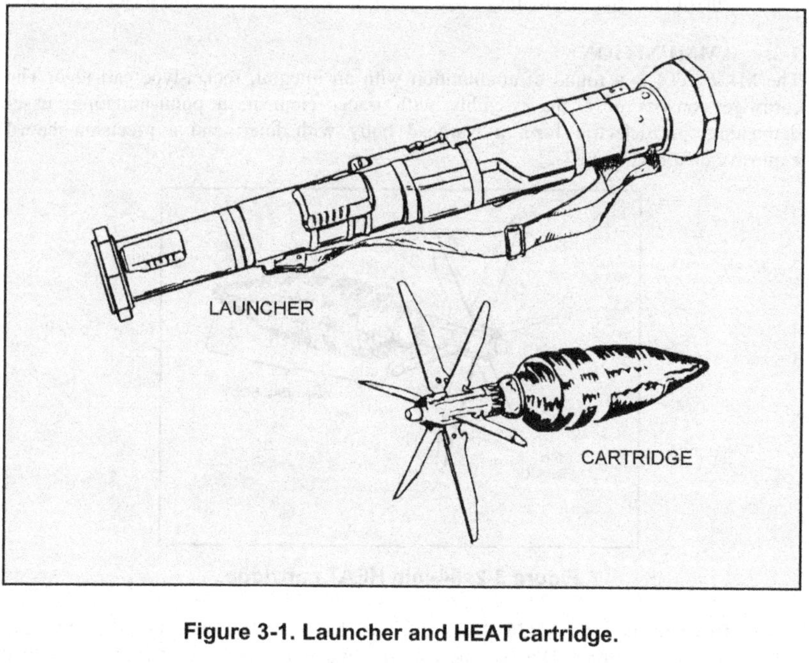

Figure 3-1. Launcher and HEAT cartridge.

3-2. TECHNICAL DATA

The following data apply to the M136 AT4:

a. **Launcher**.

Length.. 1,020 mm (40 inches)
Weight (Complete System) .... 6.7 kg (14.8 pounds)
Rear Sight.............Range indicator, graduated in 50-meter increments

b. **Rocket**.

Caliber.. 84 mm
Muzzle Velocity .... 290 mps (950 fps)
Length.. 460 mm (18 inches)
Weight .... 1.8 kg (4 pounds)
Minimum Range
  Training .... 30 meters (100 feet)
  Combat .... 10 meters (33 feet)
  Arming .... 10 meters (33 feet)
Maximum Range .... 2,100 meters (6,890 feet)
Maximum Effective Range .... 300 meters (985 feet)

3-3. AMMUNITION

The M136 AT4 is a round of ammunition with an integral, rocket-type cartridge. The cartridge consists of a fin assembly with tracer element; a point-initiating, base-detonating, piezoelectric fuze; a warhead body with liner; and a precision-shaped explosive charge (Figure 3-2).

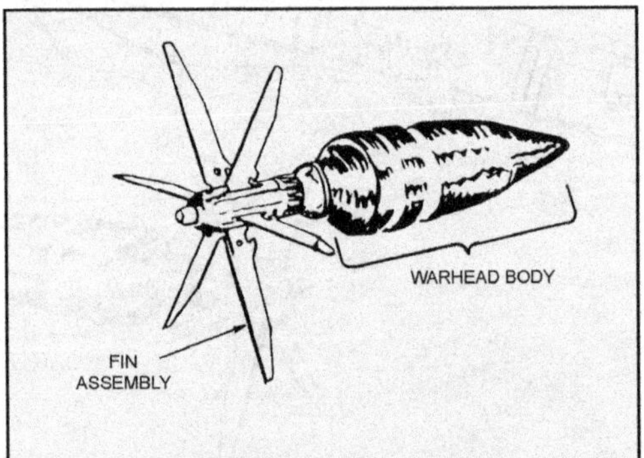

Figure 3-2. 84-mm HEAT cartridge.

a. **Description**. The M136 AT4's warhead has excellent penetration ability and lethal after-armor effects. The extremely destructive, 440 gram shaped-charge explosive penetrates more than 14 inches (35.6 cm) of armor. Warhead effects are shown in Figure 3-3.

(1) ***Impact***. The nose cone crushes; the impact sensor activates the fuze.

(2) ***Ignition***. The piezoelectric fuze element activates the electric detonator. The booster detonates, initiating the main charge.

(3) ***Penetration***. The main charge fires and forces the warhead body liner into a **directional gas jet that penetrates armor plate.**

(4) ***After-Armor Effects (Spalling)***: The projectile fragments and incendiary effects produce blinding light and destroy the interior of the target.

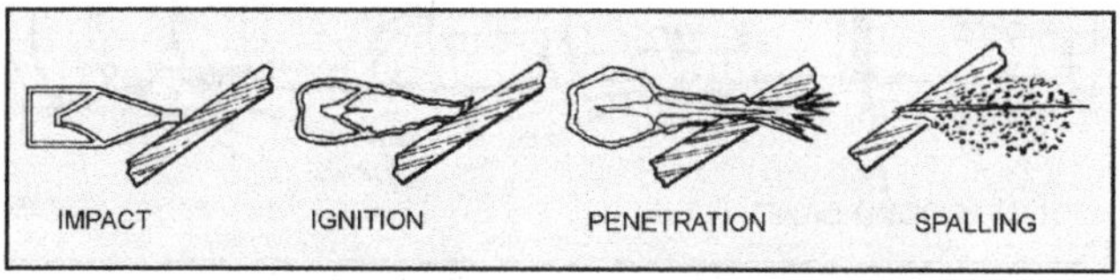

Figure 3-3. Effects of M136 AT4 warhead.

b. **Packaging**. Five M136 AT4s, each wrapped in a plastic barrier bag, are packed together in a wooden container. The containers are too heavy to stack more than four deep on the pallets (Figure 3-4).

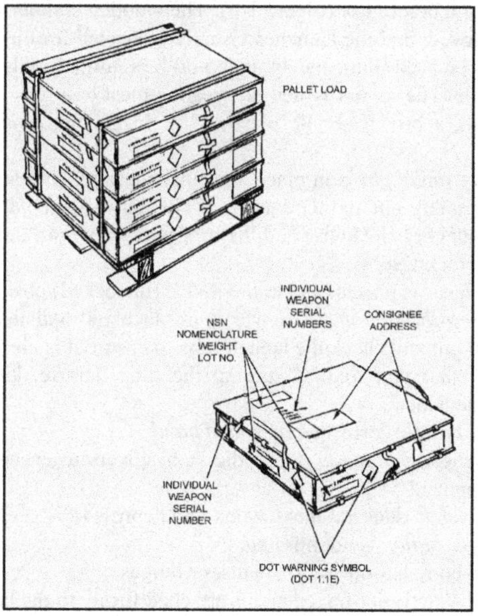

Figure 3-4. Ammunition packaging and markings.

c. **Color-Coding**. M136 AT4 launchers are marked with color-coded bands (Figure 3-5). A black with yellow band indicates an HE antiarmor round (early models had a solid black band). A gold or yellow band indicates a field handling trainer; no band indicates an M287 9-mm tracer bullet trainer (Appendix B).

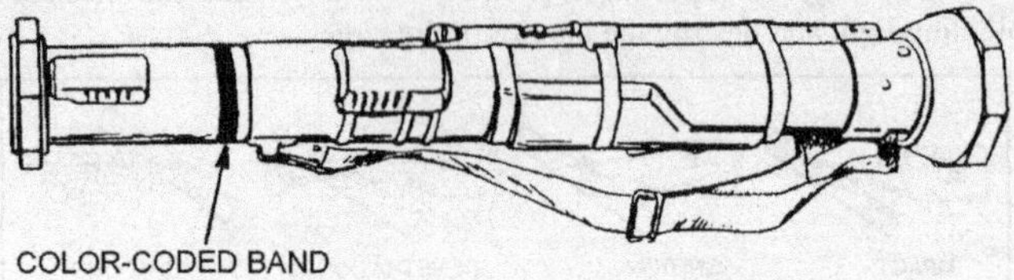

Figure 3-5. Location of color-coded band.

3-4. INSPECTION

Since the M136 AT4 is issued as a round of ammunition rather than as a weapon, the launcher is completely sealed. However, its overall condition should be inspected at the time of issue and again before use (Figure 3-6). The wooden container should be opened, the plastic bags removed, and the launcher visually inspected for obvious damage. If the M136 AT4 is not to be used immediately, it should be returned to its plastic bag and the bag resealed with tape. The soldier issued the weapon must ensure—

- The *rear seal*, a brown acrylic plastic plate inside the *venturi*, is in place and undamaged.
- The *transport safety pin* is in place and fully inserted. The lanyard is attached to the transport safety pin and the launcher. The lanyard should already be wrapped around the launcher clockwise and the transport safety pin inserted in the retainer hole counterclockwise.
- The *cocking lever* is present and in the SAFE (uncocked) position.
- The plastic *fire-through muzzle cover* is in place and undamaged. If it is torn or broken, cut it out and check the launch tube to ensure it is clear of foreign objects. Remove any that you find by turning the tube muzzle downward and gently shaking the launcher.
- The launcher has the correct *color-coded band.*
- The *sights* function properly. Open the sight covers to ensure the sights pop up and are undamaged.
- The *forward safety* does not move when you depress it.
- The *red trigger button* is not missing.
- The launcher body has no cracks, dents, or bulges.
- The *carrying sling* is not frayed and is attached firmly to the launch tube.
- The *shoulder stop* is not broken or damaged, and it unsnaps and folds down.

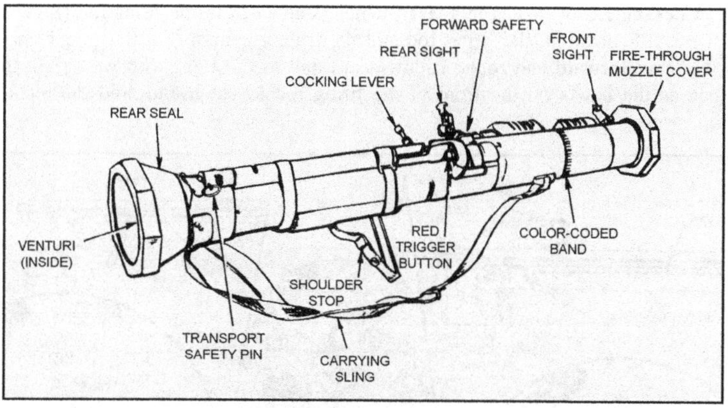

Figure 3-6. Inspection.

3-5. FIRING MECHANISM, SAFETIES, AND WEAPON FUNCTION

The function of the M136 AT4 must be discussed along with its firing mechanism and safety features. The firing mechanism is mechanical and consists of a red trigger button, an enclosed firing rod and spring, and three safety devices (Figure 3-7). The first safety device is the *transport safety* pin. The red trigger button is located between the other two, the *cocking lever* and the *forward safety*. The weapon cannot be fired until all three safeties have been disengaged.

a. **Transport Safety Pin**. This pin blocks the firing pin from striking the cartridge percussion cap. To disengage this pin, pull it outward, then release it.

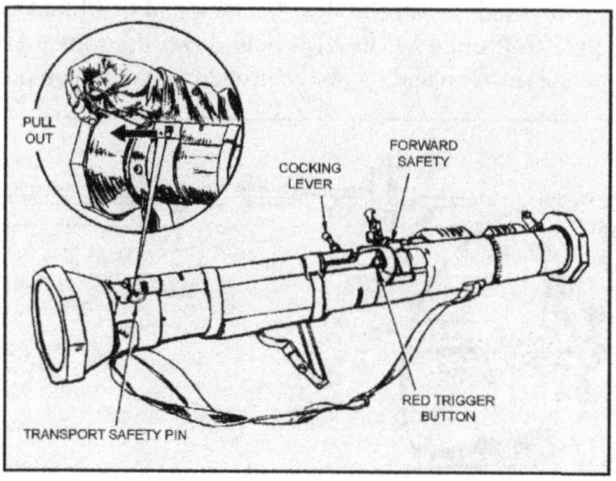

Figure 3-7. Firing mechanism and transport safety pin.

b. **Cocking Lever**. When this lever, which is attached to the firing rod (Figure 3-8), is in the SAFE position, the firing rod and the trigger cannot touch. To cock the AT4, push the lever forward and rotate it downward and to the right with your right thumb. This causes the hooks on the front of the firing rod to catch and hold the red trigger button.

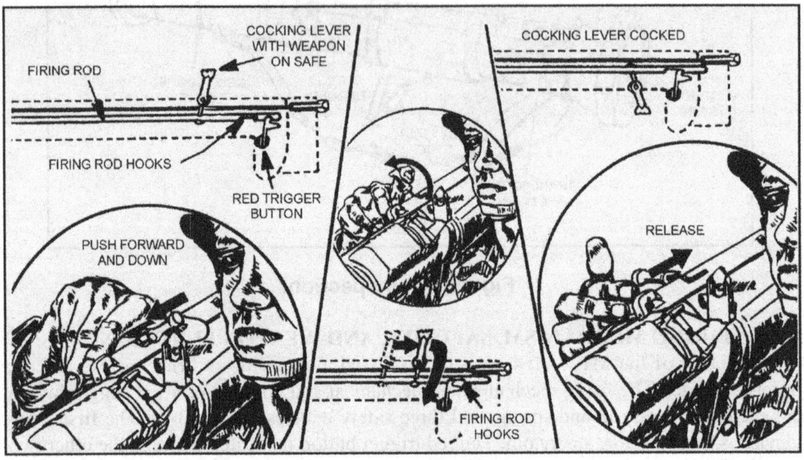

Figure 3-8. Cocking lever.

c. **Forward Safety**. This safety is on the front end of the firing mechanism (Figure 3-9) and is connected to a steel rod with a bent end that blocks the firing rod from striking the firing pin. To fire the M136 AT4, hold down the forward safety so the firing rod can strike the percussion cap and ignite the propellant when you push the trigger.

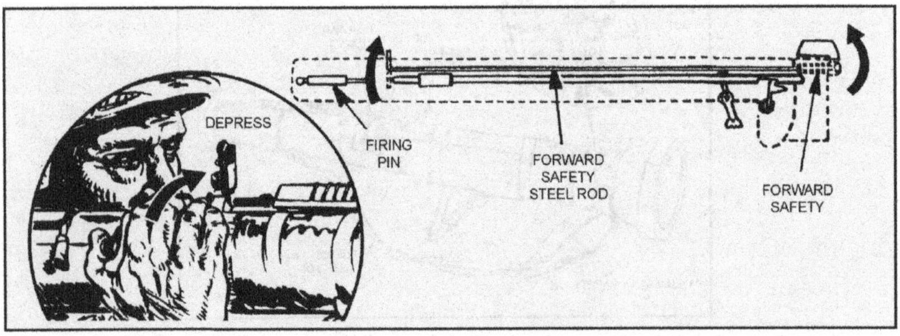

Figure 3-9. Forward safety.

3-6. SIGHTS

The fact that the AT4's front and rear sights resemble those of the M16-series rifle makes using the AT4 easier (Figure 3-10).

a. **Front Sight**. The front sight has a sight blade with a center post and left and right lead posts. A semicircular white line helps you obtain the proper sight picture. To open the front sight cover, press down on it and slide it backward until the sight pops up.

b. **Rear Sight**. The rear sight has a sight blade, range adjustment knob, range scale, 2-mm peephole for normal daylight visibility conditions, and 7-mm peephole for limited visibility conditions. To open the rear sight cover, press down on it and slide it forward until the sight pops up.

c. **Nightsight**. The AT4 can be fitted with the AN/PAQ-4C, AN/PEQ-2, or the AN/PAS-13, when used with the nightsight mounting bracket (NSN 5340-01-391-3004).

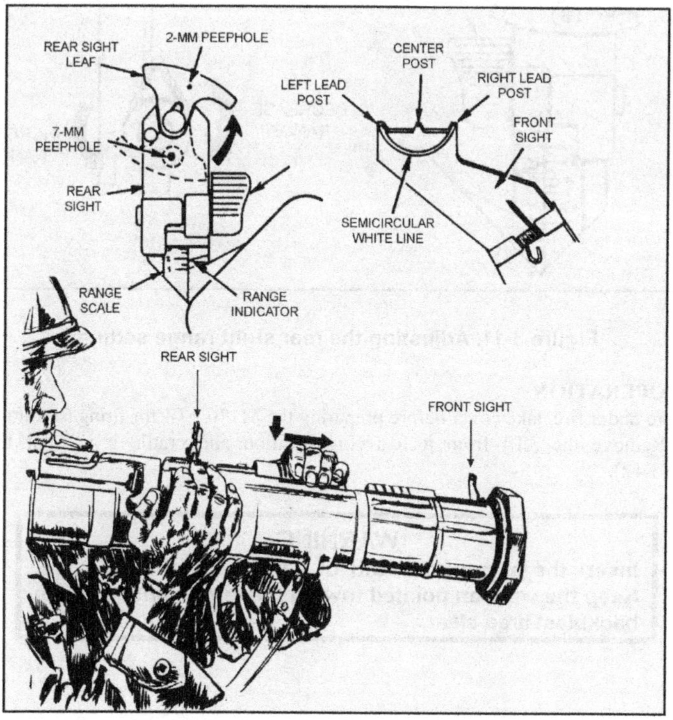

Figure 3-10. Sights.

(1) The leaf blade that covers the 7-mm peephole has its own tiny 2-mm peephole. To uncover the 7-mm peephole, pull the bottom of the leaf blade out slightly and rotate it right and up. To cover the 7-mm peephole, rotate it back down and ensure the leaf blade

is seated. The range indicator scale is indexed from 100 to 500 meters in 50-meter increments.

(2) To increase the range setting beyond 200 meters, turn the range adjustment knob clockwise, or vice versa (Figure 3-11). You must remember to reset the range to 200 meters when you close the rear sight. Otherwise, closing the sight cover will break off the rear sight.

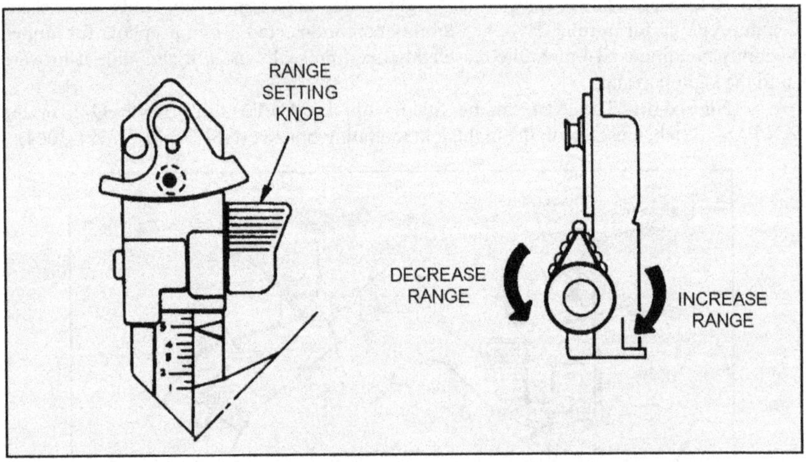

Figure 3-11. Adjusting the rear sight range setting.

3-7. OPERATION

If you are under fire, take cover before preparing the M136 AT4 for firing as follows:

a. Remove the AT4 from its carrying position and cradle it in your left arm (Figure 3-12).

> **WARNING**
> **Insert the approved brand of earplugs before you fire. Keep the weapon pointed toward the target, and keep the backblast area clear.**

Figure 3-12. Cradle position.

b. With your right hand, pull and release the transport safety pin (Figure 3-13). This pin is important—you must reinsert it if you do not fire the launcher. Therefore, unless it is attached to the launcher with a lanyard, you must keep it in a safe place.

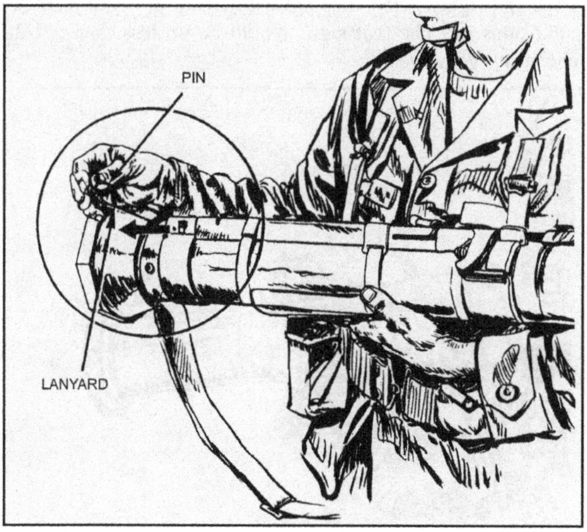

Figure 3-13. Removing the transport safety pin.

c. Unsnap, unfold, and hold the shoulder stop with your right hand (Figure 3-14).

d. Place the launcher on your right shoulder and stabilize it by grasping the sling near the launcher's muzzle with your left hand.

Figure 3-14. Unsnapping the shoulder stop.

e. With the AT4 on your right shoulder, stabilize it with your left hand and open the sights with your right hand. Press down and pull backward on the front sight cover until the front sight pops up (Figure 3-15), then press down and forward on the rear sight cover until the rear sight pops up. The rear sight should be no less than 2 1/2 inches and no more than 3 inches from your eyes.

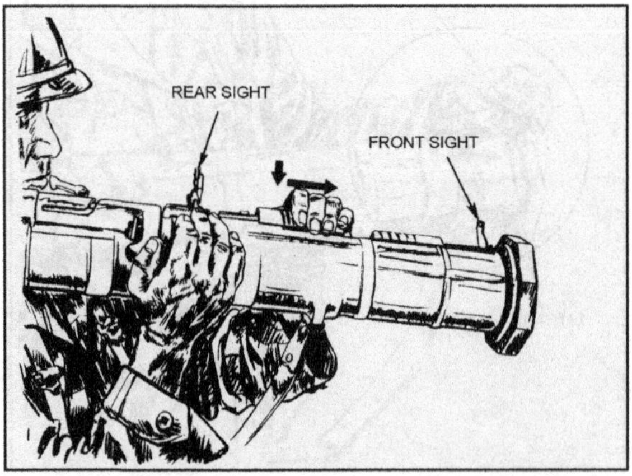

Figure 3-15. Opening and adjusting the sights.

f. Set the rear sight for the correct range to the target.

g. Check the backblast area before you cock the launcher. Then, with your right hand, unfold the cocking lever (Figure 3-16). Place your thumb under it and, with the support of your fingers in front of the firing mechanism, push it forward, rotate it downward and to the right, and let it slide backward.

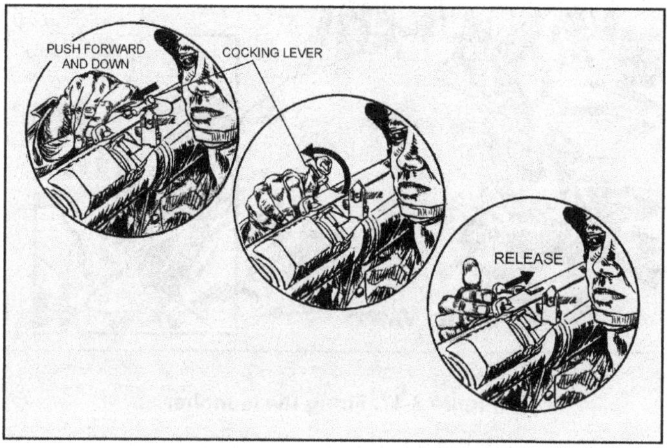

Figure 3-16. Cocking the launcher.

h. Pull back on the sling with your left hand to seat the shoulder stop firmly against your shoulder. To avoid a misfire, use the index and middle fingers on your right hand to hold the forward safety down and to the left while you fire (Figure 3-17, page 3-12).

3-8. MISFIRE PROCEDURES

A misfire is a complete failure to fire caused by a procedural or mechanical failure. Which misfire procedures should be used depends on whether the firer is in a combat or training environment.

a. **Causes.** A misfire is usually caused by one of the following factors:

The forward safety is not depressed far enough to disengage the safety.

The firing mechanism is faulty.

The propelling charge explosive train is faulty.

> **WARNING**
> **Keep your weapon pointed toward the target.**

Figure 3-17. Firing the launcher.

b. **Combat Environment**. If a misfire occurs in combat—

(1) Release the forward safety.

(2) Remove your right hand from the firing mechanism and cock the weapon again.

(3) Try to fire again. If the launcher still does not fire, maintain the same firing position and return the cocking lever to the SAFE (uncocked) position.

(4) Move the launcher from your shoulder, keeping the launcher pointed toward the enemy. Reinsert the transport safety pin.

(5) In combat, break off the sights to identify the misfired launcher. In training, however, you would not want to damage the field handling trainer (FHT), so identify the misfired launcher simply by leaving the sights up.

(6) Place the launcher on the ground, pointed toward the enemy, and use another launcher. As soon as you can, dispose of the misfired launcher IAW unit SOP.

c. **Training Environment**. If a misfire occurs on a live-fire training range, the soldier responds as follows (the trainer later disposes of the launcher IAW local SOP):

(1) Shouts "Misfire" as soon as the launcher fails to fire while maintaining the original sight picture.

(2) Releases the forward safety.

(3) Recocks the launcher: Immediately removes his right hand from the firing mechanism and pushes the cocking lever forward with the heel of his right hand until the lever locks with a loud clicking noise.

NOTE: Because performing immediate action takes so little time, you need not recheck the backblast area.

(4) Press the forward safety all the way down and try to fire again. If the launcher still fails to fire, shout misfire, release the forward safety, and move the cocking lever to the SAFE (uncocked) position. Move the launcher from your shoulder, keeping the weapon pointed toward the target and cradled in your left arm.

(5) Reinsert the transport safety pin, wait two minutes, then carefully lay the launcher on the ground, muzzle toward the target.

NOTE: Notify the local ammunition supply and issue point of any unusual occurrence, regardless of whether the weapon fires or not. Examples include excessive overpressure, recoil, or heat on your face after you have fired the weapon (caused by the propellant burning after the round leaves the muzzle).

3-9. RESTORATION TO CARRYING CONFIGURATION

If the launcher is prepared to fire, but then is not fired, it must be taken out of operation as follows:

a. Release the forward safety.

b. Push forward and to the left on the cocking lever, and let it spring back into the SAFE (uncocked) position.

c. Move the launcher from your shoulder, ensuring the muzzle is pointed in the direction of fire.

d. With the launcher cradled in your left arm, replace the transport safety pin until it is fully seated in the retainer hole.

e. To avoid breaking off the rear sight, remember to reset the range indicator to the 200-meter setting before closing the rear sight cover.

f. Lay down the sights and close their covers.

g. Snap the shoulder stop into the closed position.

h. Sling the launcher over your right shoulder and move to another location.

CHAPTER 4
MARKSMANSHIP FUNDAMENTALS

Many factors contribute to light antiarmor weapon marksmanship. Soldiers who combine these factors well, and continue to practice doing so, can retain their skills. The factors are grouped into four basic areas known as marksmanship fundamentals: steady hold, aiming procedures, breath control, and trigger manipulation. Instructions are given for right-handed firers, but the M72-series LAW can be fired from either shoulder by simply reversing the instructions.

4-1. STEADY HOLD

Maintaining a steady hold involves holding the launcher as steady as possible while sighting and firing. To maintain the proper sight picture and sight alignment until he fires, the firer must hold the launcher in a tight, comfortable position so that it becomes a natural extension of his body (Figure 4-1). With both weapons, keep your elbows close to your body to help balance the weapon and prevent you from jerking or flinching when you fire. In the case of the AT4, this reduces recoil.

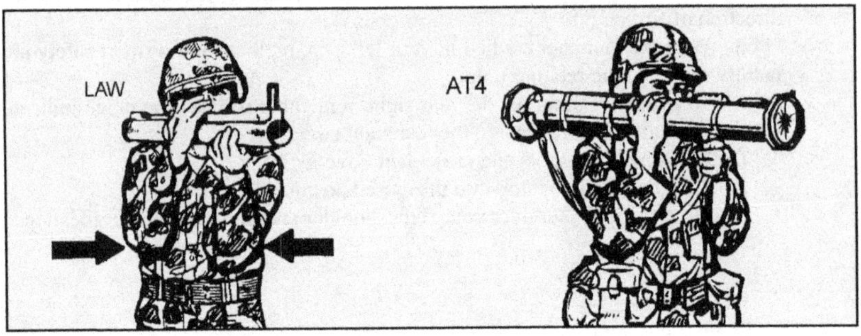

Figure 4-1. Steady hold position.

a. **M72-Series LAW**. Place your left hand, palm facing upward, under the launcher near the muzzle and grasp the launcher. Firmly pull the rear cover into your right shoulder pocket.

b. **M136 AT4**. With your left hand, grasp the carrying sling where it attaches to the launcher near the muzzle. With your right hand on the trigger mechanism, pull the shoulder stop into your right shoulder pocket.

4-2. AIMING PROCEDURES

Aiming procedures include placing the eye correctly, obtaining a sight picture, and aligning the sight. Combining these procedures is critical to correctly aiming light antiarmor weapons.

a. **Eye Placement**. Before sighting the weapon, estimate the range (Chapter 6 discusses range estimation). For the M72-series LAW, place your firing eye as close to the rear sight as is comfortable. However, for the M136 AT4, place your firing eye between 2 1/2 to 3 inches (no nearer than 2 1/2 inches) from the rear sight. This distance is necessary to prevent possible injury from the weapon's recoil and for correct sight alignment with the AT4 (Figure 4-2).

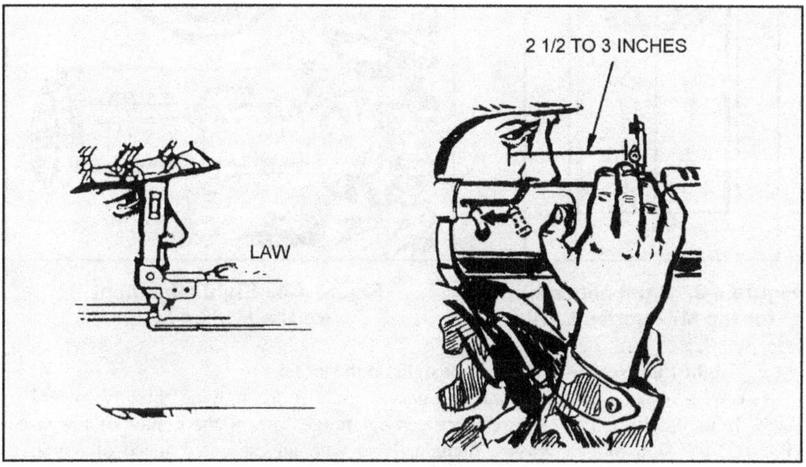

Figure 4-2. Eye placement.

> **WARNING**
> **When firing the M136 AT4, do not place your eye within 2 1/2 inches of the rear sight. The AT4s recoil could cause the rear sight to injure your firing eye.**

b. **Sight Alignment**. Align the sights correctly with the target. To do this for the M72-series LAW, position the rear sight so that your eye is near and in line with the peephole in the rear sight. Look through the peephole at the front sight reticle and place the range line that corresponds to the target's range on the target (Figure 4-3). To do this for the AT4, position the rear sight so that the white semicircle of the front sight is a hazy line around the bottom half of the rear sight opening. Position the front sight posts on the target (Figure 4-4). Align the sight by moving your head forward or backward.

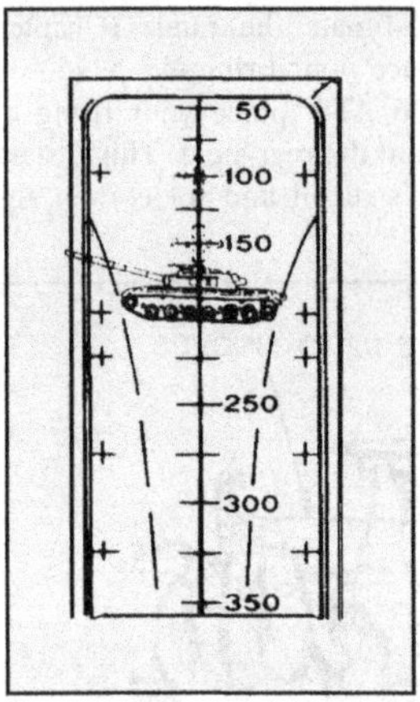

Figure 4-3. Sight alignment for the M72-series LAW.

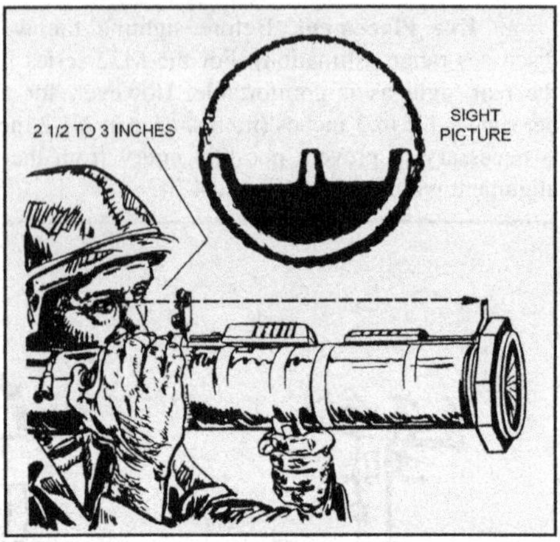

Figure 4-4. Sight alignment for the M136 AT4.

c. **Sight Picture**. Position the front sight on the target.

(1) ***M72-Series LAW***. *Stationary targets* include those moving directly toward or away from the firer. Place the correct vertical range line in the center of the target (Figure 4-5). *Slow-moving targets* include those with an estimated speed of 5 mph or slower, or those moving in an oblique direction. Place either the left or right lead cross mark on the vehicle's center of mass (Figure 4-6). *Fast-moving targets* include those with an estimated speed of more than 5 mph. Place either the left or right lead cross mark on the leading edge of the vehicle (Figure 4-7).

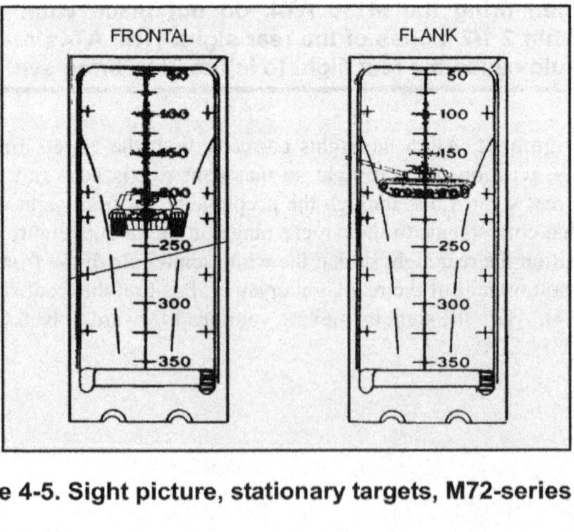

Figure 4-5. Sight picture, stationary targets, M72-series LAW.

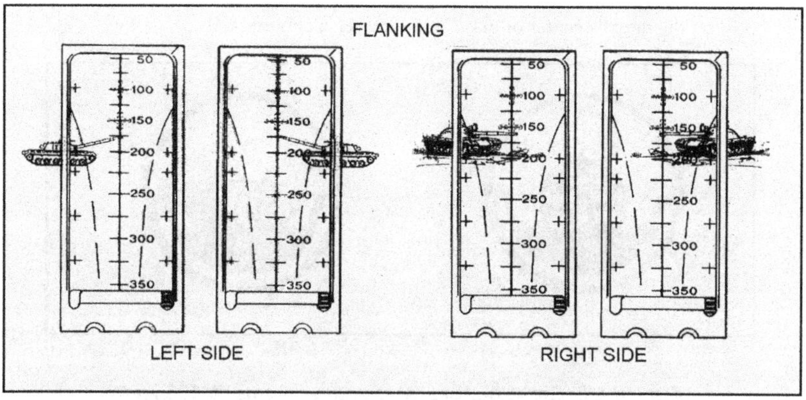

Figure 4-6. Sight picture, slow-moving targets, M72-series LAW.

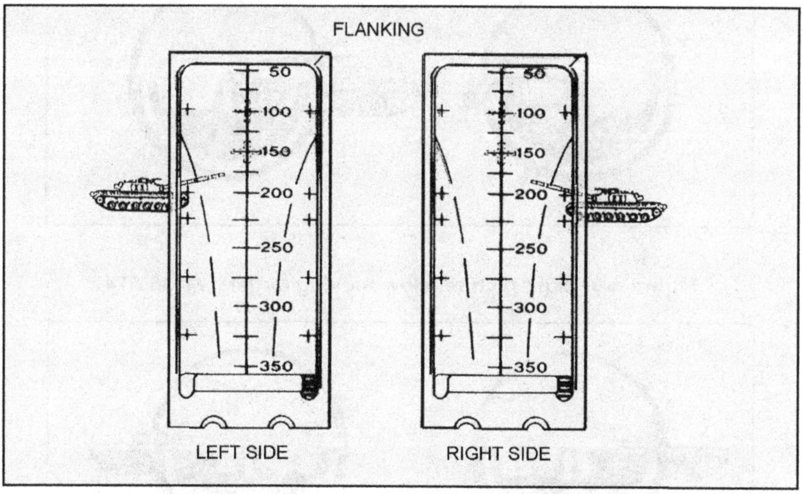

Figure 4-7. Sight picture, fast-moving targets, M72-series LAW.

(2) ***M136 AT4***. *Stationary targets* include those moving directly toward or away from the firer. Adjust the rear sight for the correct range and place the center sight post in the center of the target (Figure 4-8). *Slow-moving vehicles* are those with an estimated speed of 10 mph or less, or those moving in an oblique direction. Place the center sight post on the front or leading edge of the vehicle (Figure 4-9). *Fast-moving vehicles* are those estimated to be moving faster than 10 mph. Place either the left or right lead post on the

center of the target. For example, if the target is moving from left to right, place the left lead post on the target's center of mass, and vice versa (Figure 4-10).

Figure 4-8. Sight picture, stationary targets, M136 AT4.

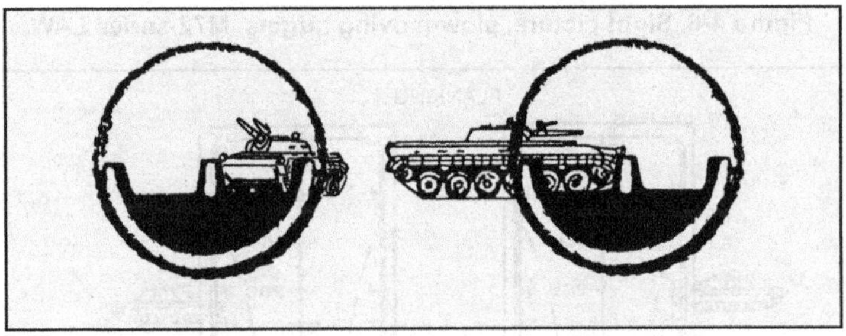

Figure 4-9. Sight picture, slow-moving targets, M136 AT4.

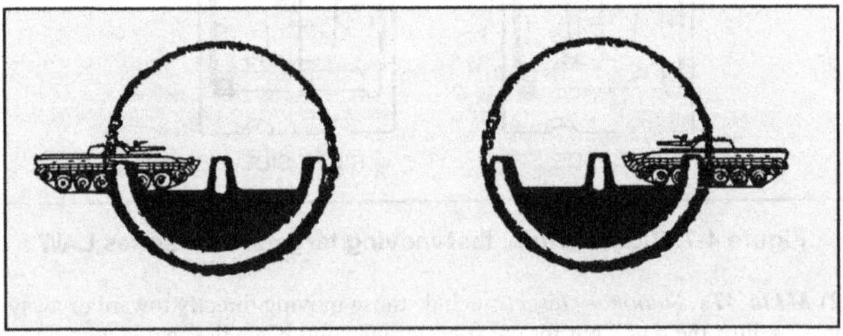

Figure 4-10. Sight picture, fast-moving targets, M136 AT4.

4-3. BREATH CONTROL

Breath control is as important when firing a light antiarmor weapon as it is when firing an individual weapon. Breathing while firing can cause a miss. To control breathing, the firer breathes deeply a couple of times, takes one last deep breath, exhales partly, holds his breath, sights, and fires.

4-4. TRIGGER MANIPULATION

Light antiarmor weapons have different types of triggers (Figure 4-11).

a. **LAW**. To fire the LAW, the firer must apply firm and steady downward pressure to the trigger with the fingers of his firing hand.

b. **AT4**. To fire the AT4, the firer must apply firm and steady forward pressure to the trigger with the thumb of his firing hand. Soldiers can practice trigger manipulation and control techniques on an expended launcher or FHT.

Figure 4-11. Trigger manipulation.

4-5. INTEGRATED ACT OF SHOOTING

Correct sight alignment is critical. Sight alignment errors increase as the range to the target increases. Therefore, maintaining the correct relationship between the rear and front sights is as important as placing the aiming point. The steps for doing this should become automatic. No matter how quickly they are done, these steps are always distinct, because the human eye can only focus at one distance and on one point at a time. The firer focuses on the front sight to obtain correct sight alignment, then places the aiming point to complete the sight picture. He shifts or adjusts the position of the launcher as necessary. The whole time he is pressing the trigger, he maintains the sight picture.

CHAPTER 5
FIRING POSITIONS

This chapter explains the basic firing positions used with light antiarmor weapons. Instructions for each are given for right-handed firers, but the M72-series LAW can be fired from either shoulder by simply reversing the instructions. Though each weapon can be fired from all four of the basic firing positions, individual physique determines exact body and hand positions. Firing from a supported position naturally increases accuracy, which improves the odds for a first-round hit or kill. Basic safety considerations are the same for all light antiarmor weapons, but additional considerations for each firing position are provided here.

5-1. STANDING POSITION
Two standing positions are used: a basic standing position and one modified for the infantry fighting position.

a. **Basic Standing Position**. Raise the launcher slightly higher than shoulder level. Execute a left face, rotate your shoulder under the launcher, and spread your feet a comfortable distance apart. Move your left foot 15 to 24 inches forward, keeping your hips level and your weight balanced on both feet. To obtain a firm, stable position, tuck both elbows tightly into your body. To track a moving target, turn your body at the waist—not with your legs. This enables you to track the target smoothly. Unless you are behind a protective barrier such as a wall, the standing position exposes you more than any other position to enemy observation and possible suppression. Differences between weapons with respect to the standing position are as follows (Figure 5-1, page 5-2):

WARNING
Always keep the launcher pointed in the direction of fire.

(1) ***M72-Series LAW***. Place your nonfiring hand about 4 inches from the front of the muzzle, with your firing hand on the rear cover. After placing the weapon on your shoulder, release the rear cover and place your firing hand on the trigger. Cup the launcher in the palm of your nonfiring hand. Position your firing eye as close to the rear sight as is comfortable.

(2) ***M136 AT4***. Grasp the sling near the launcher with your left hand and the shoulder stop with your right hand. Raise the launcher above shoulder level. After placing the launcher on your shoulder, release the shoulder stop and place your right hand on the trigger. Place your firing eye 2 1/2 to 3 inches from the rear sight.

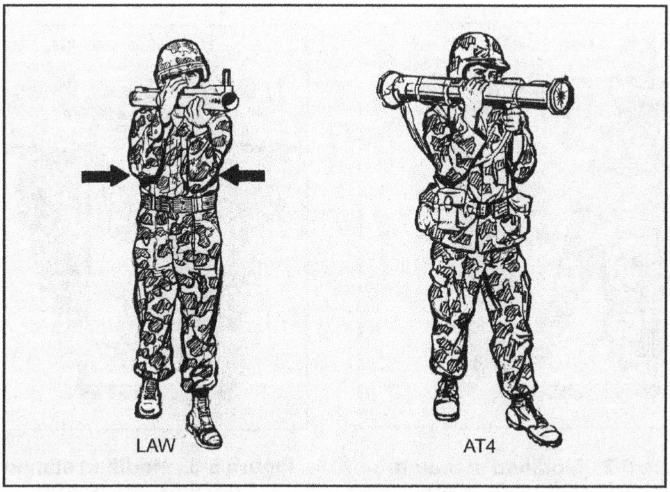

Figure 5-1. Basic standing position.

b. **Modified Standing Position**. Use this position when you occupy an infantry fighting position. Assume the basic standing position, but instead of stepping forward, lean against the back wall of the fighting position. Ensure that the venturi or rear of the weapon extends beyond the rear of the fighting position. Figure 5-2 shows the modified standing position for the M72-series LAW; Figure 5-3 shows the modified standing position for the M136 AT4. Ensure that NONE of the following are in your backblast area:

- Other soldiers.
- Other fighting positions.
- Equipment.
- Any part of your own fighting position.
- Obstructions within 5 meters.

NOTE: Leaders must ensure that light antiarmor weapons are positioned so that the backblast misses other fighting positions.

DANGER

NEVER FIRE FROM WITHIN A COMPLETELY ENCLOSED, UNVENTILATED BUNKER OR FIGHTING POSITION.

Figure 5-2. Modified standing, position, M72-series LAW.

Figure 5-3. Modified standing position, M136 AT4.

5-2. KNEELING POSITION

The basic kneeling position is the best position for tracking moving targets. The modified kneeling position is best for engaging stationary targets, since it is a supported position. However, either can be used for stationary or moving targets.

a. **Basic Kneeling Position**. Kneel from the basic standing position onto your right knee, keeping your left thigh parallel to the ground. Rotate your lower right leg 90 degrees to the left. (This removes your right foot from exposure to the backblast.) Keep your right thigh and back straight and perpendicular to the ground. Point your left foot in the direction of fire and tuck your elbows in to your sides. Though this is not a supported position, it should be a firm, stable one. Figure 5-4 shows the basic kneeling positions for the LAW and AT4.

b. **Modified Kneeling Position**. From the basic kneeling position, sit back on your right heel. Place the back of your upper left arm on your left knee, making sure you do not have bone-to-bone contact between your left elbow and left knee. Keep your right elbow tucked in close to your right side. Use any protective barriers available. Figure 5-5 shows the modified kneeling positions for the LAW and AT4.

Figure 5-4. Basic kneeling positions.

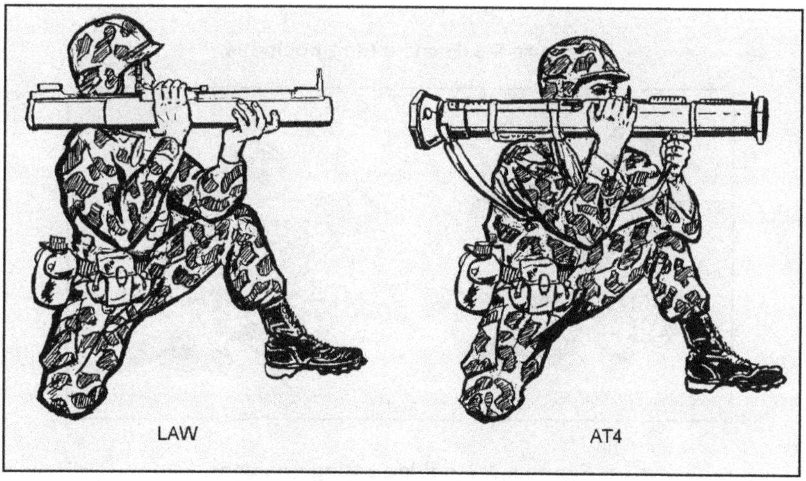

Figure 5-5. Modified kneeling positions.

5-3. SITTING POSITION

The sitting position is the most stable firing position. In this position, the arms are placed on the legs for support. Depending on his physique, the firer can use either of two versions of the sitting position. Either is suitable for engaging stationary targets.

a. **Basic Sitting Position**. Sit on your buttocks while facing the target, and spread your feet a comfortable distance apart. Lean forward and place the backs of your upper arms on your knees, avoiding bone-to-bone contact. Figure 5-6 shows the basic sitting positions for the LAW and AT4.

b. **Modified Sitting Position**. From the basic sitting position, cross your ankles for added support. Raise or lower your knees to adjust for elevation on the target. Figure 5-7 shows the modified sitting positions for the LAW and AT4.

Figure 5-6. Basic sitting positions.

Figure 5-7. Modified sitting positions.

5-4. PRONE POSITION

The prone position is the most dangerous position due to its proximity to the ground. Ideally, the ground should slope downward from the rear of the launcher. This slope reduces the effects of the backblast.

a. Lie on your stomach with your body at a 90-degree angle to direction of fire, and with your body and legs to the left of the direction of fire.

b. *Ensure that neither the body nor the legs are in the backblast area.*

c. Unlike other firing positions, this one prevents you from placing the launcher on your right shoulder. Instead, you must hold the launcher in place against your upper right arm. For stability, apply extra pressure on the firing mechanism with your right hand. The prone position is the *least* stable of all firing positions. You must practice it often to become confident using it. Figure 5-8 shows the prone positions for the LAW and AT4.

DANGER

FAILURE TO MAINTAIN A 90-DEGREE ANGLE FROM THE DIRECTION OF FIRE COULD CAUSE INJURY OR DEATH TO THE FIRER.

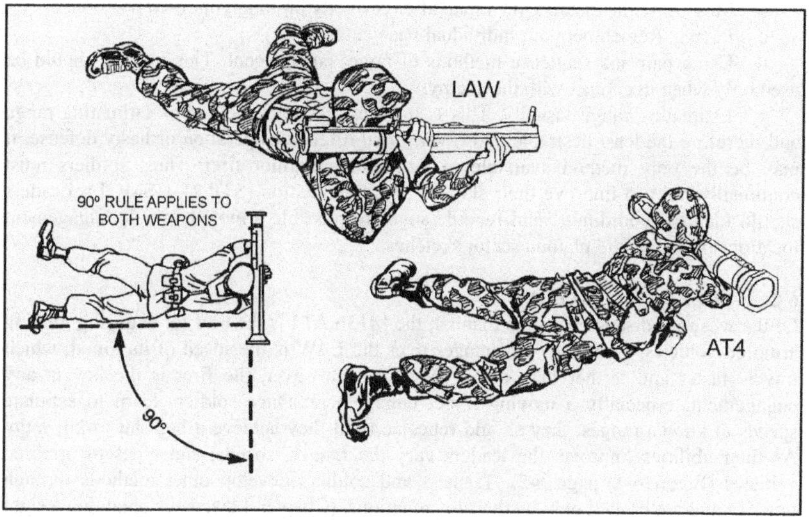

Figure 5-8. Prone positions.

CHAPTER 6

COMBAT TECHNIQUES

This chapter discusses employment techniques for light antiarmor weapons, all of which require at least basic gunnery skills. Techniques that require advanced skills are identified as such.

6-1. RANGE ESTIMATION

A firer who can accurately estimate the range to the target has a better chance of hitting it, regardless of the weapon used. Common methods of estimating range are listed below from the most to the least accurate. The tactical situation determines the method to be used:

a. Using range finders.

b. Measuring the distance on a map after correctly plotting your own position.

c. Pacing. Remember your individual pace count.

d. Using pair and sequence methods of target engagement. This method should be used only when in contact with the enemy.

e. Estimating range visually. This is the least accurate method of estimating range and therefore the least desirable. However, in an offensive operation or hasty defense, it may be the only method available to the light antiarmor firer. Thus, soldiers must continually train to improve their skill at visual estimation (STP 21-1-SMCT). Leaders should identify, coordinate, and record ranges to possible armored vehicle engagement locations on squad and platoon sector sketches.

6-2. SPEED ESTIMATION

Of the weapons discussed in this manual, the M136 AT4 is the best for engaging moving armored vehicles. One of its advantages over the LAW is the speed of its round, which travels faster and farther than the LAW round. However, the firer is the key in any engagement, especially a moving target engagement. Once soldiers learn to estimate speeds at known ranges, they should rehearse until they achieve a high hit-to-kill ratio. As their abilities improve, the leaders vary the ranges, speeds, and types of armored vehicles (Figure 6-1, page 6-2). Trainers and soldiers develop other methods through practice and are limited only by their imaginations. (Chapter 4 discusses obtaining a sight picture in detail.) Estimate how far the vehicle travels in 1 second:

a. Start when the front end of the vehicle passes the object.

b. Count, "One thousand and one" (takes about one second).

c. If *more* than half of the vehicle passes the object, estimate it as a *fast-moving* vehicle (10 mph or faster). If *less* than half of the vehicle passes the object, estimate it as a *slow-moving vehicle* (less than 10 mph).

Figure 6-1. Speed estimation.

6-3. ARMORED VEHICLE WEAKNESSES

Armored vehicles usually have their heaviest armor in front, because they are designed mainly for offensive operations against other armored vehicles (Figure 6-2). All vehicles are vulnerable to repeated hits on their flanks and rear, though the flank offers the largest possible target. Firers should always aim center of mass to increase the probability of a hit. The older the vehicle model, the less protection it has against antiarmor weapons. Consequently, newer versions may use bolt-on (applique) armor to improve their survivability. Some vehicles are equipped with reactive armor, which consists of metal plates and plastic explosives. Reactive armor usually covers the forward-facing portions and sides of the vehicle and can defeat shaped-charge weapons such as the LAW and AT4. When reactive armor detonates, it disperses metal fragments to 200 meters. The M72-series LAW and the M136 AT4 cause only a small entry hole in an armored vehicle target, though some fragmentation or spall may occur.

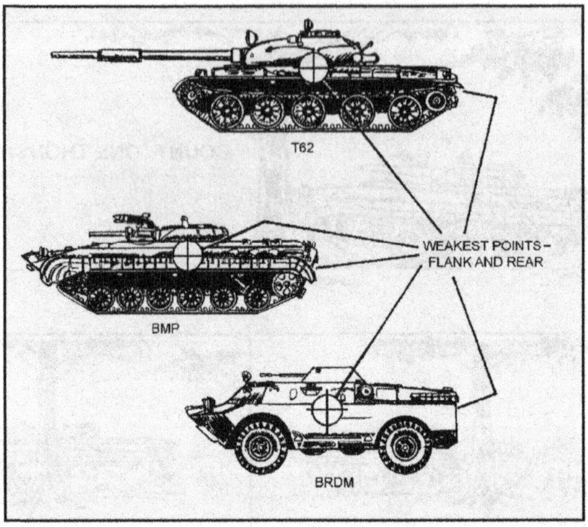

Figure 6-2. Armored vehicle weak points.

a. Natural or man-made obstacles can be used to force the armored vehicle to slow, stop, or change direction. This pause enables the firer to achieve a first-round hit. If he does not achieve a catastrophic kill on the first round, he or another firer must be ready to engage the target vehicle immediately with another round.

b. An armored vehicle without close protection (dismounted infantry) in woods, MOUT, or other restrictive terrain is vulnerable to close attack. This type of attack is most likely to originate from well-armed infantry-type teams organized into armor-killer teams. (Noninfantry units may also be required to perform this mission.) Skilled firers from these teams should engage the suspension or engine compartment of vehicles that have applique or reactive armor. When an armored vehicle is buttoned up—all hatches are closed and personnel are inside the vehicle—the crew cannot see well enough to protect itself from close attacks or attacks from the flanks or rear. The personnel inside cannot see anything within 10 meters of the vehicle, and they cannot shoot at anything (using their main guns) within 20 meters. The white area in Figure 6-3, page 6-4 shows the most favorable direction of attack when the turret is facing to the front; the gray area shows the vehicle's principal direction of fire and observation when the turret is facing to the front.

c. Armored vehicle kills are classified according to the level of damage achieved (Table 6-1).

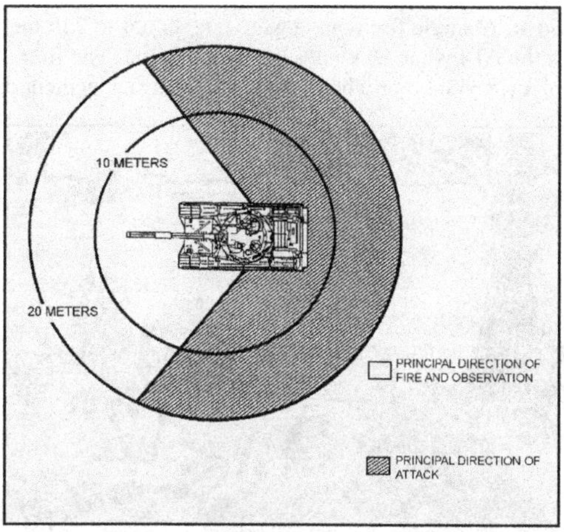

Figure 6-3. Limited visibility of armored vehicles.

TYPE OF KILL	PART OF VEHICLE DAMAGED OR DESTROYED	CAPABILITY AFTER KILL
Mobility Kill	Suspension (track, wheels, or road wheels) or power train (engine or transmission) has been damaged.	Vehicle cannot move, but it can still return fire.
Firepower Kill	Main armament has been disabled.	Vehicle can still move, so it can get away.
Catastrophic Kill	Ammunition or fuel storage section has been hit by more than one round.	Vehicle completely destroyed.

Table 6-1. Armored vehicle kills.

6-4. METHODS OF ENGAGEMENT

The four engagement methods include single, sequence, pair, and volley firing. The leader evaluates the situation on the ground to determine which of these methods to use. Regardless of whether they are used singly or in combination, communications are needed as well. The methods of engagement are rehearsed IAW unit SOP.

a. **Single Firing**. A single soldier with one light antiarmor weapon may engage an armored vehicle, but this is not the preferred method of engagement. Several light antiarmor weapons are required to kill an armored vehicle. A single firer firing one round must hit a vital part of the target to damage it at all (Figure 6-4).

(1) ***Range Not Known***. A single firer should engage only targets within 200 meters when he does not know the actual range. The probability that he hits a target beyond 200 meters with a single round is small.

(2) ***Range Known***. A single firer can engage targets out to 225 meters with the LAW or 300 meters with the AT4 when he knows the actual range. The firer should do this only when he has a flank or rear shot, or when he has no other engagement option.

Figure 6-4. Single firing.

b. **Sequence Firing**. A single firer, equipped with two or more light antiarmor weapons prepared for firing, engages the target. After engaging with the first round and observing the impact, the firer adjusts his point of aim, engages with another round, and so on until he destroys the target or runs out of rounds (Figure 6-5).

Figure 6-5. Sequence firing.

c. **Pair Firing**. Two or more firers, equipped with two or more light antiarmor weapons prepared for firing, engage a single target. Before firing, the first firer informs the others of the estimated speed and distance to the target. If the impact of his round proves his estimate to be correct, the other firers engage the target until it is destroyed. If the impact of the round proves his estimate to be incorrect, the second firer informs the others of his own estimate, then he engages the target. This continues until the target is destroyed or all rounds are expended (Figure 6-6).

Figure 6-6. Pair firing.

d. **Volley Firing**. Two or more firers can engage a single target when the range is known. These firers engage the target at one time on a prearranged signal such as a command, whistle, booby trap, mine, or TRP. For the best method of engagement, use the light antiarmor weapon, because it places the most possible rounds on one target at one time, increasing the possibility of a kill (Figure 6-7).

Figure 6-7. Volley firing.

e. **Communications**. Leaders control all unit fire and communicate this information to the entire unit IAW unit SOP. Light antiarmor weapons firers must know—

- Designated firers.
- Target priority.
- Method of engagement.
- Range and lead to target (if known).
- Command or signal to fire.
- Command or signal to cease fire.

6-5. ENGAGEMENT OF FIELD FORTIFICATIONS AND BUILDINGS

Light antiarmor weapons have little effect against field fortifications and buildings. Soldiers should not expect to severely damage targets with these weapons. However, if the alternatives shown in Table 6-2, page 6-8, are used, soldiers may be able to gain a temporary advantage.

6-6. ENGAGEMENT OF OTHER VEHICLES

The M72-series LAW proves more effective against light vehicles; the M136 AT4 proves more effective against armored vehicles. Nonarmored vehicles, such as trucks, cars, and boats, are considered *soft targets*. Firing along their length offers the greatest chance of a kill, because this type of shot is most likely to hit their engine block or fuel tank.

AIM POINT	EFFECT WHEN WEAPON IS FIRED AT AIM POINT	RECOMMENDED FIRING TECHNIQUE
BUNKER OR FIGHTING POSITION		
Firing Port or Aperture	Rounds fired into firing ports or apertures are wasted: rounds detonate inside rear of position, causing little structural damage to the position or to the equipment or personnel within, unless they are hit directly. The AT4 produces less effect than the LAW.	Coordinate fire: Fire *light antiarmor weapons* at a point 6 to 12 inches from the front edge of the firing ports in the berm. Fire *small arms* at the bunker or position to prevent personnel within from returning light antiarmor fire.
Berm	Firing at the berm causes the round to detonate outside the fighting position or inside the berm, creating only a small hole in the berm, dust, or minor structural damage to the position, but no damage to personnel or equipment unless they are hit directly. The AT4 produces less effect than the LAW.	
Window	The round may travel completely through the structure before detonating; if not, it creates dust and causes minor structural damage to the rear wall, but little damage to personnel or equipment, unless they are hit directly. The AT4 produces *less* effect than the LAW.	Fire 6 to 12 inches from the sides or bottom of a window. Light antiarmor rounds explode on contact with brick or concrete, creating an opening whose size is determined by the type of round used.
Wall	The round detonates on contact, creating dust and causing a small hole and minor structural damage, but little damage to personnel or equipment, unless they are hit directly. The LAW may be used to create a loophole, which is a hole large enough to throw hand grenades through. The AT4 produces *less* effect than the LAW.	
Corner	Corners are reinforced and thus harder to penetrate than other parts of a wall. Any light antiarmor round will detonate sooner on a corner than on a less dense surface. Detonation should occur in the targeted room, creating dust and causing *overpressure*, which can temporarily incapacitate personnel inside the structure near the point of detonation. The AT4 causes *more* overpressure than the LAW.	

Table 6-2. Effects of light antiarmor weapons on field fortifications or bunkers.

6-7. LIMITED VISIBILITY ENGAGEMENTS

Limited visibility engagements can be conducted using various night vision devices or with artificial illumination. However, when NVDs or artificial illumination is used, limited visibility can reduce the maximum effective range for light antiarmor weapons by at least one-third. To avoid fratricide, leaders must ensure all designated light antiarmor weapon firers are trained to use their weapons in limited visibility.

a. **Night Vision Device**. Before a NVD can be used with the M136/AT4, it must be removed from its designated weapon (M249 machine gun or automatic weapon, or M60 machine gun) and the M136/AT4 mounting bracket must be attached. Appendix F describes the various NVDs that can be used, it also gives information for mounting, boresighting, and zeroing procedures for each NVD.

b. **Artificial Illumination**. If artificial illumination is used during a limited visibility engagement, it should be placed above and slightly beyond the target. However, the ability to identify and engage targets is even less with artificial illumination than with NVDs.

6-8. ENGAGEMENT IN NBC CONDITIONS

Wearing a protective mask limits the firer's ability to sight the weapon. Wearing NBC gloves limits his ability to manipulate the firing mechanism.

a. **Sighting the Weapon**. Sighting while wearing the protective mask may require rotating the weapon slightly counterclockwise. The mask also makes determining the location, identity, and engageability of targets more difficult.

b. **Firing the Weapon**. Practice manipulating the firing mechanism while wearing NBC gloves.

NOTE: When live firing either a light antiarmor weapon or its subcaliber trainer, aim within range firing limits.

DANGER

THE M136 AT4 IS NOT RATED SAFE. IN TRAINING, NEVER FIRE IT FROM AN ENCLOSURE; IN COMBAT, DO SO ONLY WHEN NO OTHER POSITION IS AVAILABLE. THE OVERPRESSURE AND BLAST COMBINED CREATE CONDITIONS THAT CAN KILL YOU. IF THEY DON'T KILL YOU, THEY CAN INJURE YOU SERIOUSLY OR DEAFEN YOU TEMPORARILY OR PERMANENTLY.

6-9. ENGAGEMENT FROM AN ENCLOSURE

Firing from an enclosure creates unique hazards. As such, before positioning soldiers in enclosures (in combat only), leaders must consider several factors that affect safety. Only in combat, *when no other tactical option exists*, should the M136 AT4 be fired from an enclosure. If it must be employed this way, the enclosure must meet the following minimum requirements. The M72-series LAW has been rated as safe for use from an enclosure but, again, only when the enclosure meets the following minimum requirements:

a. **Construction.** The building must be sturdily constructed to reduce the structural damage that would occur in a weakly constructed enclosure such as one made of wood or stucco.

b. **Size of Enclosure.** Minimum measurements for the enclosure are as follows:

- AT4—minimum room size 17 x 24 feet.
- LAW—minimum room size 12 x 15 feet.
- Both—minimum ceiling height 8 feet.

c. **Ventilation to the Rear and Sides.** To allow for the backblast, at least 20 square feet of ventilation—such as a standard 3-foot by 7-foot doorway—must be provided directly behind the firer. More doors and windows are removed beside and behind the position to increase ventilation and reduce overpressure, noise, and blast effects. Without sufficient ventilation, the blast would weaken or collapse the walls. On the front wall, windows and doors are reinforced rather than removed, because removing would draw attention to the position. Reinforcing the windows also helps protect the firer from enemy direct-fire weapons.

d. **Objects and Debris.** All objects and debris are removed from the rear of the weapon, because the backblast causes them to fly around the room injuring personnel.

e. **Muzzle Clearance.** Muzzle Clearance must be at least 6 inches.

f. **Weapon Clearance.** Properly positioning the weapons within the enclosure is vital to the safety and survival of all personnel in the enclosure. The weapons should be positioned so that no walls are within 5 meters to the rear or side of the weapon.

g. **Personnel Positions.** If any other soldiers must be present, then they must remain forward of the rear of the launcher and avoid standing in corners or near walls. If possible, they should construct reinforced positions that fit the previous criteria and that can protect them in case the building collapses.

WARNING

To avoid injuring the eardrums, soldiers must wear the approved brand of ear protection.

6-10. ENGAGEMENT BEYOND MAXIMUM EFFECTIVE RANGE (M136 AT4 ONLY)

A skilled M136 AT4 firer can engage targets beyond the weapon's maximum effective range of 300 meters, up to 550 meters. Beyond 550 meters, the firer must aim higher than center of mass and apply additional lead for moving targets. Commanders must realize that accuracy is reduced at these ranges. Also, firing at these ranges identifies the firing position to the enemy.

6-11. OFFENSIVE OPERATIONS

All elements, even those with other organic antiarmor weapons, use light antiarmor weapons. Light antiarmor weapons can influence the action in an attack, so units should routinely stock them beforehand. They are most useful against lightly armored vehicles. They can also be used against soft targets, such as bunkers, field fortifications, automobiles, and trucks, but their shaped-charge warheads have less effect on these than on armored targets. Unless personnel, ordnance, or flammable material on or inside them are hit, soft targets can normally continue to fight after being attacked by light antiarmor weapons. Due to their relatively short range, light antiarmor weapons should be placed throughout the attacking force. They support the maneuver by providing a base of fire, and they enable the assaulting force to engage in close antiarmor combat.

6-12. DEFENSIVE OPERATIONS

Whether or not other organic antiarmor weapons are available, light antiarmor weapons are an asset in the squad and platoon defensive plan.

a. Light antiarmor weapons are employed with interlocking fires to provide mutual support. Dispersion allows leaders the flexibility to place flank, rear, and oblique fires on targets. This procedure increases the survivability of the firers as well as the probability of achieving kills. Leaders must select positions that avoid fratricide from antiarmor backblast and short rounds (Appendix A).

b. The squad and platoon leader's sector sketch identifies TRPs and primary areas of possible engagement. The sector sketch also identifies possible avenues of approach for enemy armored vehicles. Each light antiarmor weapon shares a sector of fire with the primary small-arms weapon assigned to its position. However, because it is neither a primary nor crew-served weapon, the light antiarmor weapon does not require a separate range card. When assigning sectors of fire, squad and platoon leaders inform firers of all possible target areas, TRPs, and prearranged signals (Figure 6-8). (FMs 7-7, 7-7J, and 7-8 provide more information about the squad and platoon sector sketch.) Leaders provide each designated position with two or more light antiarmor weapons.

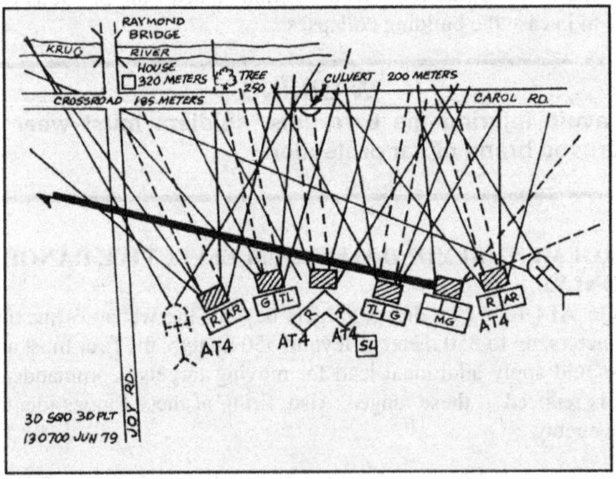

Figure 6-8. Squad sector sketch.

6-13. OTHER TACTICAL OPERATIONS

The weight of the light antiarmor weapon suits it well for combat patrols and rear area operations. (FM 7-8 provides more information about these subjects.)

a. **Combat Patrols**. Light antiarmor weapons are used on combat patrols to destroy enemy equipment, installations, and key points, and to harass enemy forces. The two types of combat patrols are—

(1) ***Ambushes (Antiarmor)***. The commander can employ armor-killer teams either during limited visibility or when cover, concealment, and withdrawal routes are available. The key to ambushing armored or other vehicles is to choose terrain that restricts their maneuverability and fields of fire, but that allows friendly forces to engage the vehicles from the flank and rear. Soldiers can emplace antipersonnel mines before the ambush so that dismounting enemy soldiers deploy into them. Though light antiarmor weapons can be used independently, they are normally used in support of designated organic antiarmor weapons such as the Dragon. Volley firing light antiarmor weapons increases the probability of a quick kill (Figure 6-9). Security teams stop the enemy from escaping.

Figure 6-9. Antiarmor ambushes.

(2) ***Raids***. Using light antiarmor weapons on raids increases target options. Light antiarmor weapons can also be used to defend against armored vehicles.

b. **Rear Area Operations**. Rear area operations usually consist of either: defending unit trains and tactical operation centers, or patrolling rear areas.

(1) ***Defense of Tactical Operations Center and Unit Trains***. The mobility of modern vehicles makes the TOC and unit trains prime targets. Light antiarmor weapons are used to engage vehicles threatening the unit during the defense of the TOC. Soldiers who use light antiarmor weapons in this type of operation normally perform noninfantry-type roles. Unit leaders designate soldiers to carry light antiarmor weapons and ensure that these soldiers receive light antiarmor weapons training.

(2) ***Patrols of Rear Areas***. Rear area patrols are security patrols conducted by designated infantry or MP units. These patrols can react to any threat they encounter in the rear area. Rear area patrols use the light antiarmor weapon in a hasty point defense at a roadblock, intersection, or strongpoint.

CHAPTER 7

TRAIN-THE-TRAINER AND UNIT TRAINING PROGRAMS

During initial-entry training, soldiers receive only an orientation for each of the light antiarmor weapons. When they arrive in a unit, soldiers should train to Skill Level 1 proficiency on all assigned weapons and equipment. Each unit should have an ongoing unit training program to allow soldiers to work together to attain and sustain their skills. This program integrates individual and collective tasks. Operating a unit training program requires the unit to also maintain a train-the-trainer program. Both programs are planned and conducted IAW the commander's METL.

7-1. TRAINING STRATEGY

Training strategy integrates resources into a year-round program to train the individual and collective skills needed to perform the wartime mission. Thus, units are trained to fight and win on the battlefield. The training strategy includes institutional and unit training.

a. **Institutional Training**. TRADOC schools conduct institutional training. These courses train critical individual and leader skills required to succeed in combat. Students are provided the latest doctrine and taught current technical skills associated with the light antiarmor weapons available within the Army. Soldiers train at several institutional schools throughout their career (see the appropriate programs of instruction [POIs]) (Table 7-1).

ENLISTED	Initial-Entry Training
	Primary Leadership Development Course
	Basic Noncommissioned Officer Course
	Advanced Noncommissioned Officer Course
OFFICERS	Precommission Training
	Officer Basic Course
	Officer Advanced Course

Table 7-1. Institutional training.

b. **Unit Training**. Unit training consists of sustainment training in individual and collective skills. Commanders manage their resources (devices, simulators, simulations, ranges, and ammunition) carefully to best develop individual and leader skills. Through the use of exercises, (drills, STXs, and qualification tables), the unit develops skills and integrates individuals into a cohesive crew or squad. Table 7-2 provides a training guide for building a light antiarmor weapons sustainment training plan (see also Appendixes C and E).

7-2. UNIT TRAINING PROGRAM

Every unit commander, including those commanding CS and CSS units, should develop a program for light antiarmor training. Some units have more than one type of light antiarmor weapon that requires sustainment training. The commander considers preparation and training for both collective and individual tasks. Trainers analyze their collective tasks and the individual tasks that support them. Individual tasks must be integrated into collective training and rehearsals. The units should try to train light antiarmor weapons at the same time and on the same range (Appendix D). Otherwise, these weapons are trained in alternating quarters of the year.

a. **Sustainment Training Requirements.** Sustainment training is recommended for each soldier. Active Component units should conduct sustainment training quarterly and fire the appropriate qualification tables semiannually, depending on the STRAC TRC level. Reserve Component units should conduct sustainment training semiannually and fire the appropriate instructional or qualification tables annually. All units should fire advanced instructional tables semiannually. When possible, they should conduct live fire annually IAW the current STRAC manual. (Appendix E provides all the necessary firing tables.) Commanders should ensure that designated gunners complete unit sustainment training and live fire tactical light antiarmor weapons sometime during that quarter. *Training requirements* are divided into four tracks, A through D, based on ammunition allocations (Table 7-3). Table 7-4 provides an example *training frequency chart.* This example is based on the AT4. The same tasks apply to all light antiarmor weapons, but procedures and standards may vary. Tactical weapons are allocated IAW the current standards in training commission (STRAC) manual.

1	Designated soldiers should train as often as needed to maintain proficiency.
2	Performance evaluations should be administered to designated soldiers quarterly, if resources allow.
3	Designated soldiers should fire the qualification tables semiannually for verification on designated weapons.
4	Designated soldiers qualify semiannually on their designated weapons. Qualification consists of a hands-on performance evaluation and qualification firing tables.
5	Designated soldiers should be trained in advanced *instructional* techniques as often as the unit training program allows. This training helps to maintain their target engagement skills.
6	All soldiers not designated as light antiarmor weapons gunners should train on these weapons to the extent allowed by resources.

Table 7-2. Sustainment training plan for light antiarmor weapons.

TRACK	DETERMINES ANNUAL AMMUNITION ALLOCATION FOR:
A	Each Active Component designated firer.
B	Track deleted
C	Each Reserve Component designated firer.
D	Each Reserve Component training division instructor.

Table 7-3. Training requirements.

TRAINING OBJECTIVE	TASK	FREQUENCY		APPROXIMATE TIME
		AC	RC	
NA	Conduct an orientation safety briefing (as appropriate).	Q	A	10 min
1	Inspect an M136 AT4 for serviceability.	Q	A	30 min
2	Prepare an M136 AT4 for firing.	Q	A	45 min
3	Demonstrate correct M136 AT4 firing positions.	Q	A	45 min
4	Estimate range.	Q	A	60 min
5	Determine the correct M136 AT4 sight picture.	Q	A	30 min
6	Perform M136 AT4 misfire procedures.	Q	A	45 min
7	Return the M136 AT4 to the carrying position.	Q	A	45 min
8	Engage targets with an M136 AT4.	S	B	3 1/2 hr
KEY: **Q = quarterly (every 90 days)** **S = semiannually (twice a year)** **A = annually (once a year)** **B = biennially (every two years)**				

Table 7-4. Example training frequency chart.

b. **Gunnery Training Requirements**. All designated soldiers should be trained to standard IAW the combat techniques in Chapter 6 and the performance evaluations in Appendix C.

c. **Collective Training**. Individual tasks must be integrated into collective training and rehearsals. To accomplish this, commanders analyze the collective tasks from their unit METL and the individual tasks that support the collective tasks. The commander determines the collective tasks that must be practiced by analyzing the mission outlines in the appropriate mission training plan (MTP). The mission outlines describe the collective tasks required to execute each ARTEP mission. Leaders determine the individual tasks that support the collective tasks by referring to the mission task matrix in the appropriate MTP.

(1) Leaders use the training standards given in the appropriate manual for the individual tasks. Noninfantry units can use FMs 7-7, 7-7J, 7-8, 7-10, and 7-20 to identify infantry-type missions. Some of the tasks identified may include engaging armored vehicles. When planning training, the trainer incorporates the training devices and aids for light antiarmor weapons into the unit's training. These devices and aids add realism and allow trainer to properly evaluate the task being conducted. He must also identify deficiencies and correct them by retraining soldiers to use the selected weapons and their related training devices correctly.

(2) The MILES Viper is a force-on-force trainer for light antiarmor weapons. MILES trainers are *not* gunnery trainers and are not used for sustainment training. They are used in all force-on-force training. Table 7-5 provides an example MILES training program.

SEQUENCE	FREQUENCY (TIMES PER YEAR)	
	ACTIVE	RESERVE
Squad, platoon, or company FTX or STX	4	1
Battalion FTX	2	0
Battalion ARTEP	2	1
Total MILES training events per year	8	2

Table 7-5. MILES training program.

7-3. TRAIN-THE-TRAINER PROGRAM

The success of the unit's light antiarmor training program depends on trainers who are well-trained, competent, and prepared to train soldiers.

a. **Objectives.** The train-the-trainer program must teach trainers to do the following:

- Evaluate the unit's training weaknesses.
- Plan training.
- Set up, operate, and maintain training equipment.
- Conduct training.
- Coach firers in gunnery training.
- Evaluate firers' training weaknesses.
- Correct firers' training weaknesses.
- Instill confidence in the firers trained.
- Maintain training records.

b. **Mission-Essential Task List.** The commander must carefully examine his unit's wartime mission and develop a METL. He must then organize the unit's sustainment training program to support the METL. The light antiarmor sustainment training plan consists of quarterly, semiannual, and annual training events. Between these events, the unit provides additional light antiarmor weapons training. The sustainment plan should prepare the unit for the next round of weapons gunnery while also supporting the unit METL.

c. **Trainer Assessment**. Trainers within a unit are normally team, squad, and section leaders and platoon sergeants. Before becoming weapons trainers, they must be assessed carefully and any shortcomings corrected. The commander chooses a method of assessing the trainers that ensures that their abilities are accurately evaluated. Then, with the assistance of the platoon leaders and platoon sergeants, the commander performs this assessment.

d. **Command Benefits**. Regular and aggressive train-the-trainer programs provide valuable benefits that gradually spread through the unit. These benefits include a base of expertise, proficiency, and esprit de corps. Maintaining this base is easier than recreating it.

e. **Training Tasks**. Successful trainers know how to operate the training devices for the light antiarmor weapons assigned to their units. Based on the unit's METL, the commander designates the appropriate light antiarmor weapon for specified missions. Appendix C provides the tasks, conditions, and standards for each weapon discussed in this manual. The trainers must know the appropriate combat techniques for employing these weapons. They must also perform the following:

- Conduct an orientation safety briefing.
- Inspect the weapons for serviceability.
- Prepare the weapons for firing.
- Demonstrate the correct firing positions.
- Estimate range.
- Determine the correct sight picture.
- Perform the correct combat and training misfire procedures.
- Return the weapon to the carrying configuration.

APPENDIX A
SAFETY

This appendix provides safety procedures for both training and combat.

A-1. TRAINING SAFETY

Training safety for light antiarmor weapons applies to all types of ranges and training sites.

a. **Backblast**. Light antiarmor weapons have little or no recoil, because the propellant gasses escape to the rear of the weapon. This backblast can damage equipment or seriously injure personnel who are too close to the rear of the launcher. When operating temperatures fall below freezing (0NC or 32NF), all backblast areas and safety zones double.

(1) ***M72-series LAW***. This is an open-chambered weapon, so it has no recoil. The launcher's total backblast area extends 40 meters (44 yards) to the rear (Figure A-1) and is divided into two zones. During training, both should be marked off limits:

(a) Danger zone. All personnel, equipment, and flammable material must be clear of this area.

(b) Caution zone. The weapon's backblast may throw loose objects to the rear. Therefore, personnel must also stay clear of this area.

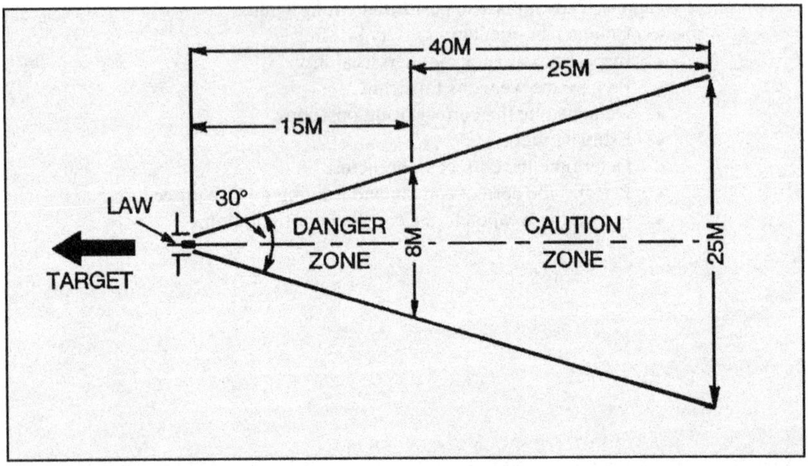

Figure A-1. M72-series LAW backblast area.

(2) ***M136 AT4***. Because the AT4 has a closed chamber, it has a slight recoil. When the weapon is fired, pressure builds up in the chamber and ruptures the baseplate. Once the baseplate ruptures, gasses exit from the back of the weapon, forming the backblast. The total backblast area extends 100 meters to the rear of the launcher in a 90-degree fan (Figure A-2, page A-2). During training, the *entire* backblast area should be marked off limits and kept clear of personnel and equipment. In combat, personnel may remain to the rear of the

weapon, but *only* if they are in protected positions. Backblast heat or flying debris could injure the firer and other soldiers anywhere in the backblast area. This weapon must not be fired from an enclosure or in front of a barrier (TM 9-1315-886-12). It may be fired from an enclosure in combat *only when* no other tactical option exists.

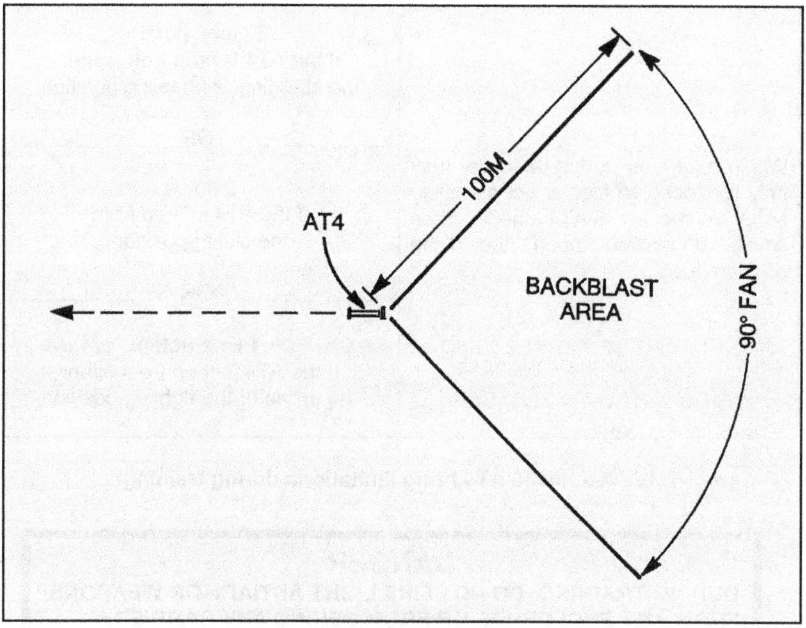

Figure A-2. M136 AT4 backblast area.

b. **Firing Limitations**. Trainers and soldiers must observe the backblast danger zones as well as other firing limitations. A light antiarmor weapon can be fired from the standard infantry fighting position. However, if this position is chosen, trainers/soldiers must consider other factors that increase the accuracy and reduce the danger to friendly soldiers: Rear of launcher extends beyond the back wall of the position, no overhead cover extends into the backblast area, and at least 1/2 meter (18 inches) clearance is around the weapon.

c. **Range Procedures for Light Antiarmor Weapons**. For all the light antiarmor weapons discussed in this manual, these procedures must be followed:

(1) The whole backblast area must be well marked and must be clear of all personnel, equipment, and flammable materials.

(2) No one enters the area behind the firing line or forward of the rear safety line without permission from the range OIC or safety NCO.

(3) Weapons are stored to prevent exposure to the sun or extreme temperatures.

d. **Additional Range Procedures for the M136 AT4**. The extreme overpressure and noise created when this weapon is fired require special precautions during training. Not only must firers and others present wear earplugs, but they may only use the approved brand—no other brand provides sufficient protection. To reduce these hazards, the range firing limitations shown in Table A-1 must be followed.

Within a 24-hour period, a soldier may only fire, observe fire, or act as safety NCO for the M136 AT4 one to three times, depending upon the firing position used	3 times (total) if the AT4 is fired from either the standing or kneeling position OR 2 times if the AT4 is fired from the sitting position OR 1 time (total) if the AT4 is fired from either the prone or the fighting position

Table A-1. M136 AT4 firing limitations during training.

DANGER

DURING TRAINING, DO NOT FIRE LIGHT ANTIARMOR WEAPONS FROM ANY ENCLOSURE OR FROM BEHIND ANY BARRIER.

A-2. COMBAT SAFETY

Combat safety rules and procedures include all those that apply to training plus the following:

a. **Fighting Position**. Both the M72-series LAW and the M136 AT4 can be fired from the standard infantry fighting position. However, to increase accuracy and reduce the danger to friendly soldiers, the area to the rear of the firing position must have no walls, large trees, or other obstructions within 5 meters (5 1/2 yards) to deflect the backblast onto the firer or into the position.

DANGER

ENSURE THE AREA TO THE REAR OF THE FIRING POSITION HAS NO WALLS, LARGE TREES, OR OTHER OBSTRUCTIONS WITIN 5 METERS (5 1/2 YARDS). OBSTRUCTIONS DEFLECT BACKBLAST ONTO THE FIRER OR INTO THE POSITION, INJURING OR KILLING THE FIRER AND ANY OTHER SOLDIER OCCUPYING THE FIRING POSITION.

(1) In the *individual infantry fighting position,* the soldier must lean against the rear wall and ensure that the venturi or the rear of the weapon protrudes past the rear of the position. The firer's elbows are not supported.

(2) In the *two-soldier infantry fighting position,* the nonfiring soldier must remain clear of the backblast area. These positions should be constructed and sited so that none are located in another's backblast danger zone.

(3) A *modified firing position* may be constructed to the side of the two-soldier fighting position. Firing from a modified position reduces the possibility of injury to the firer or another soldier in the fighting position, but it offers the firer some protection from enemy return fire.

b. **Temperature Limitations**. No light antiarmor weapon should be fired when its temperature exceeds operating limits. When a weapon temperature is below freezing (0 degrees Centigrade or 32 degrees Fahrenheit), overpressure increases and prevents some of the propellant from burning. Then, as the projectile leaves the launch tube, portions of the remaining propellant could be directed onto the firer's face or upper body. Consequently, in cold weather, the firer should wear a protective mask. On the other hand, when the weapon temperature is above 60 degrees Centigrade (140 degrees Fahrenheit), propellants and warhead compounds can destabilize, so, whether they are located at training sites or fighting positions, weapons should be stored in a shaded area in hot weather. Finally, weapons must not be fired when their temperatures exceed safe operating limits: -40 degrees to 60 degrees Centigrade (-40 degrees to +140 degrees Fahrenheit).

c. **Minimum Target Engagement Ranges**. In combat, no targets will be engaged within the 10-meter minimum arming range. During training, no targets will be engaged within 30 meters.

d. **Overhead Fire**. Light antiarmor weapons must not be fired over the heads of friendly soldiers, unless the soldiers have adequate protection against direct impact or other hazards.

APPENDIX B

TRAINING DEVICES AND AIDS

Training devices and training aids enable soldiers to learn as much as they can about a weapon before they try the real thing. This procedure saves money and time, and it prevents injuries. ***Training devices*** *both look and behave like the real weapon, but are cheaper and safer to practice with.* ***Training aids*** *include anything else used to help soldiers learn to use a weapon. Trainers should use their imaginations and invent or adapt other training aids from available resources.*

B-1. LAW SUBCALIBER TRAINING DEVICE

The only M72-series LAW training device designed for individual use is a subcaliber trainer (Figure B-1). This device, referred to as the M190 subcaliber launcher, is made by adding an M190 subcaliber conversion kit (NSN 1340-00-420-7999) (Figure B-2) to an expended M72-series LAW launcher. It is used to fire the M73 subcaliber (35-mm) rocket. This 36-mm rocket is shorter and lighter than the LAW's 66-mm tactical rocket. It simulates the tactical rocket's smoke and flight trajectory but with less noise and backblast. The expended launcher is likely to wear out before the subcaliber inner tube does. When this happens, the inner tube can be transferred to another expended launcher. The M190 subcaliber launcher can be used in all training phases, from a fixed firing line to simulated tactical situations such as a squad live-fire exercise.

NOTE: The local Training Support Center (TSC) can install the conversion kits, but DOD Regulation 5100.76-M requires that all sensitive conventional arms, ammunitions, and explosives, including expended launchers, be carefully controlled.

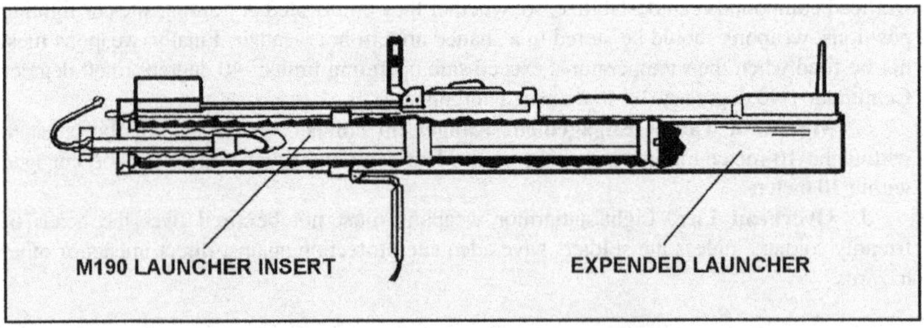

Figure B-1. LAW subcaliber trainer.

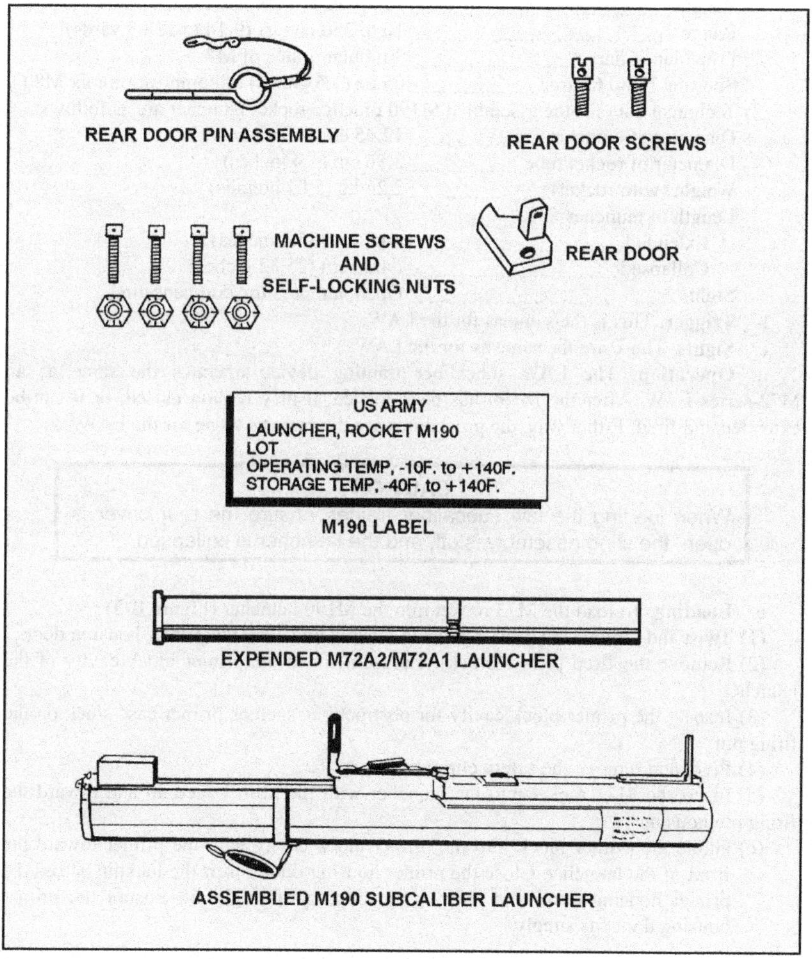

Figure B-2. M190 subcaliber conversion kit.

a. **Technical Data.**

(1) Technical data for the M73 35-mm subcaliber practice rocket are as follows:

Length .... 22.48 cm (8.86 inches)
Diameter .... 3.51 cm (1.38 inches)
Weight .... 154 grams (5.39 ounces)
Range .... 10 to 250 meters (9.14 to 228.5 yards)
Propellant Charge .... 3 tubular grains of M7
Spotting Head Charge .... 05 oz (1.5 grams) of composition mix M80

(2) Technical data for the assembled M190 practice rocket launcher are as follows:

Diameter of launcher .... 12.45 cm (4.9 inches)
Diameter of rocket tube .... 3.56 cm (1.4 inches)
Weight (with rocket) .... 2.26 kg (5.02 pounds)
Length of launcher
  Extended .... 89.9 cm (35.4 inches)
  Collapsed .... 64.26 cm (25.32 inches)
Sights .... Open, temperature compensating

b. **Trigger**. This is the same as for the LAW.

c. **Sights**. These are the same as for the LAW.

d. **Operation**. The LAW subcaliber training device operates the same as an M72-series LAW. After the M190 has been loaded, it may remain closed, or it can be extended and fired. Either way, the procedures are the same as those for the LAW.

> **CAUTION**
> When loading the law subcaliber trainer, ensure the rear cover is open, the sling assembly is off, and the launcher is collapsed.

e. **Loading**. To load the M73 rocket into the M190 launcher (Figure B-3)—

(1) Twist and remove the primer housing lockpin and open the primer housing door.

(2) Remove the fired primer block (if present) from the primer block cavity of the launcher.

(3) Inspect the primer block cavity for obstructions such as primer case stuck on the firing pin.

(4) Pivot and remove the safety clip from the rocket.

(5) Insert the M73 rocket into the launcher with the primer face up and toward the firing pin housing.

(6) Guide the primer block into the primer block cavity with the primer toward the front of the launcher. Close the primer housing door, insert the lockpin across the primer housing door, and twist to ensure a spring-tight fit. Ensure the primer housing door fits snugly.

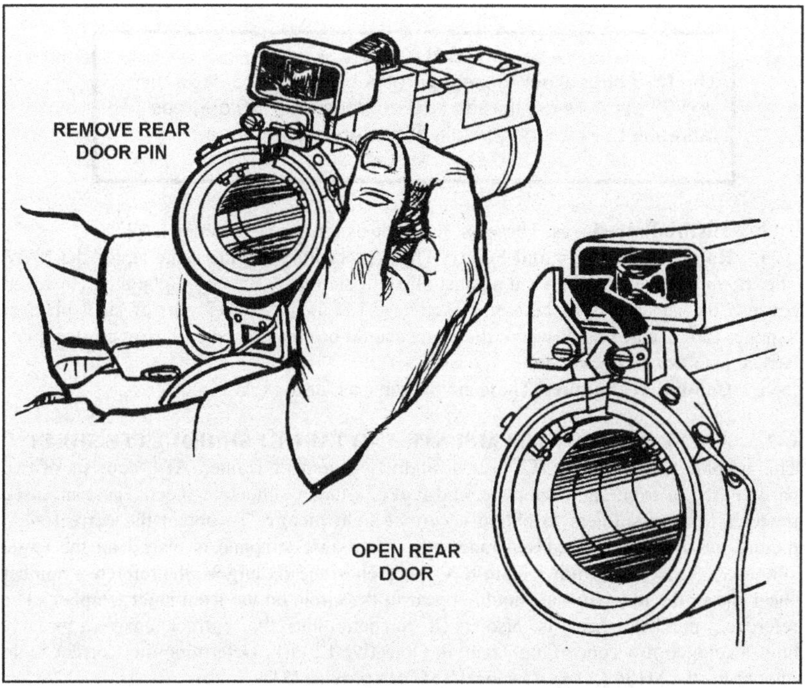

Figure B-3. Loading the LAW subcaliber launcher.

f. **Unloading**. To unload the M73 rocket from the M190 launcher—

(1) Return the arming handle to SAFE.

(2) Remove the weapon from your shoulder, keeping the weapon pointed downrange.

(3) Partly collapse the launcher, and remove the primer housing lockpin.

(4) Open the primary housing door, and remove the primer from its housing.

(5) Remove the rocket.

(6) Replace the safety clip on the rocket.

(7) Place the rocket in its original container.

g. **Function**. The M73 rocket launches the same as tactical rocket, but when the rocket head strikes the target, it sets off the spotting head that produces a flash, noise, and white smoke.

> **CAUTION**
> The M73 subcaliber rocket operates between –23 degrees and 60 degrees centigrade (-10 degrees and 140 degrees fahrenheit). Fire it only within this temperature range.

h. **Misfire Procedures**. These are the same as for the M-72 series LAW.

i. **Range Procedures and Safety**. These procedures are the same as for the LAW. The training device can be used against all solid stationary or moving targets. However, because the subcaliber rocket can penetrate 0.124 inches (0.315 cm) of steel plate or 8 inches (20.32 cm) of soft wood, the target should be constructed of 3/16-inch steel plate backed by 3/4-inch plywood.

j. **Combat Techniques**. These are the same as for the LAW.

B-2. AT4 FRONT SIGHT TEMPLATE AND TARGET SILHOUETTE SHEET

This graphic training aid (GTA 7-2-5), Sight Engagement Trainer, AT4, consists of two parts: Part 1, a front sight template, and Part 2, a target silhouette sheet. These are used together to train soldiers to obtain a correct sight picture. To obtain the correct sight picture for the target speed and range, the front sight template is placed on the target silhouette sheet as shown in Figure B-4. For each of the six targets, the reference number sheet shows the number that should appear in the circle on the front sight template. The reference number sheet is also used to determine the correct answer for any unit-developed test concerning Training Objective 12 task, Determine the Correct Sight Picture for the M136 (AT4). The local TSC can provide GTA 7-2-5.

Figure B-4. Using GTA-7-2-5 with target silhouette sheet and front sight template.

B-3. AT4 FIELD HANDLING TRAINER

The AT4 field handling trainer (FHT) is an inert device made from an expended AT4 tactical launcher. It simulates the weight, balance, characteristics, and operation of the launcher. The FHT is marked with either a gold or yellow 1-inch band between the front and rear sights, and with the word "DUMMY" in 1-inch letters on the side of the launch tube. The local TSC can convert expended launchers to training devices, but DOD Regulation 5100.76-M requires that these items be carefully controlled. The FHT is appropriate for several training situations that do not require the M287 tracer trainer, which prevents unnecessary damage to the tracer trainer. The FHT is used in marksmanship training, FTXs, and institutional training to help soldiers learn how to do the following:

- Inspect the AT4 for serviceability.
- Prepare an AT4 for firing.
- Demonstrate the correct AT4 firing position.
- Perform misfire procedures.
- Return an AT4 to carrying position.

B-4. AT4 SUBCALIBER TRACER TRAINER

The M287 subcaliber tracer trainer (Figure B-5) uses the 9-mm M939 training practice-tracer (TP-T) cartridge. When loaded, this trainer simulates the AT4 in weight, balance, and operation. The velocity and trajectory of its ammunition match that of the AT4's HEAT cartridge, but the M287 produces less noise, backblast, and overpressure. The M287 subcaliber tracer trainer is used in place of the AT4 in training.

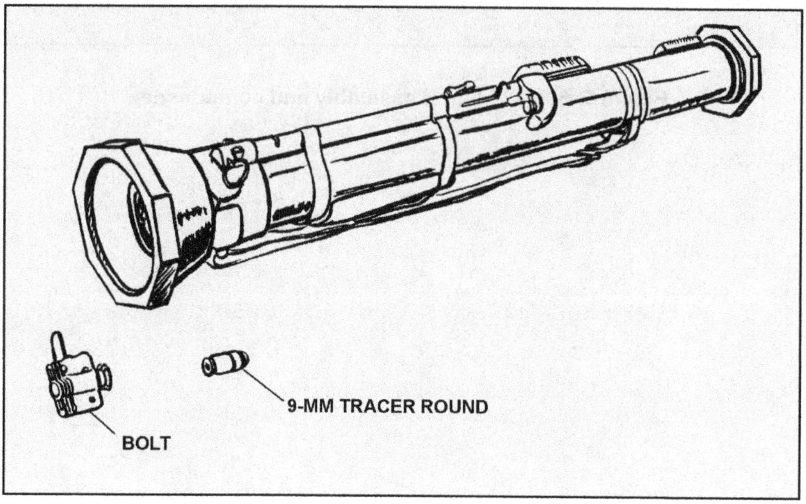

Figure B-5. M287 subcaliber tracer trainer.

a. **Description**. The M287 tracer trainer is a specially constructed AT4. Unlike the LAW subcaliber trainer, it is *not* made from an expended launcher, so it has its own NSN. It is designed to accept a special rifle barrel that fires a reduced-load 9-mm cartridge.

(1) The M287 may be fired at stationary or moving targets. Before it can be fired at a target tank occupied by personnel, the parts of the tank that could suffer damage must be shielded. The local TSC can provide specifications for modifying tanks to be used as targets for the M287 tracer trainer.

(2) Unlike the M136 AT4 itself and the FHT, the M287 has no band between the front and rear sights. The trainer has a 9-mm submachine gun barrel, a breach assembly, and a bolt (Figure B-6). The bolt is easily removed to load the 9-mm cartridges and to inspect the barrel for obstructions.

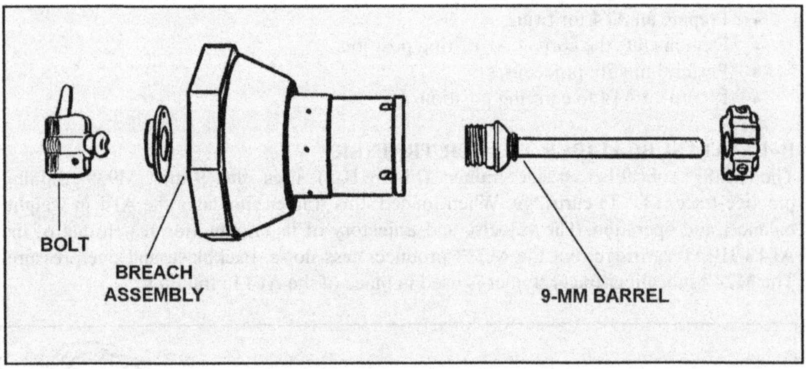

Figure B-6. M287 barrel assembly and components.

(3) If the M287 tracer trainer is damaged, the unit armorer can replace its complete firing mechanism, firing rod, bolt, sights, sling, and 9-mm barrel assembly (Figure B-7).

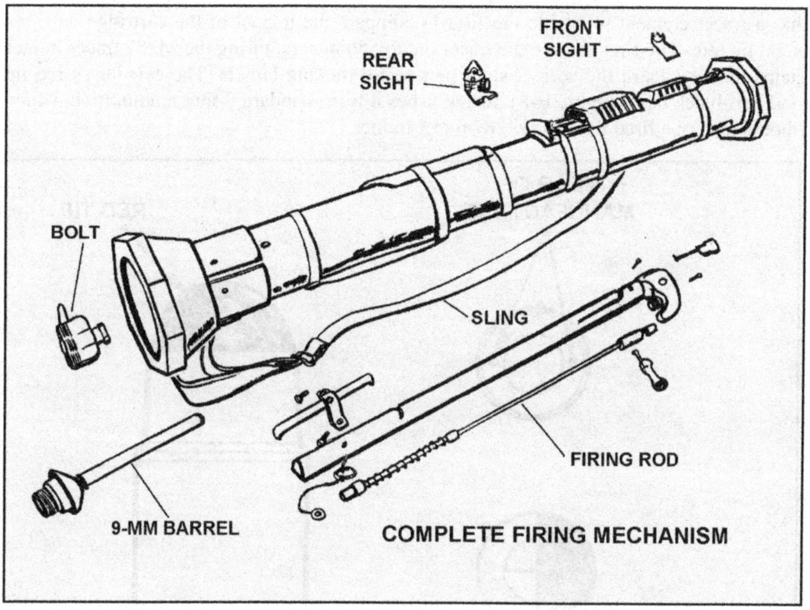

Figure B-7. Replaceable parts for the M287 tracer trainer.

b. **Technical Data**. The following technical data apply to the M287 subcaliber tracer trainer:

Length	1,020 mm (40 inches)
Weight	7 kg (15 pounds)
Action	Mechanical
Sights:	
Front	Three-post system
Rear	Adjustable range, with two peepholes: 2-mm peephole for daylight and 7-mm peephole for limited visibility
Operating temperature	10 degrees to 27 degrees Centigrade
	(0 degrees to 100 degrees Fahrenheit)
Muzzle velocity	300 mps
	(984 fps)
Caliber	9 mm

c. **Ammunition**. The M939 9-mm tracer cartridge has a lighter powder charge than a standard 9-mm bullet. The lighter charge enables the cartridge to closely duplicate the trajectory of the AT4 tactical round at ranges out to 700 meters. The M939 cartridge also has a tracer element to enable the firer to compare the impact of the cartridge with the sight picture. The firer can see the tracer out to 550 meters. Firing the M287 tracer trainer helps the firer learn the correct sight picture for moving targets. The cartridge's red tip and half-black base (Figure B-8) distinguishes it from standard 9-mm ammunition, which should never be fired from the M287 tracer trainer.

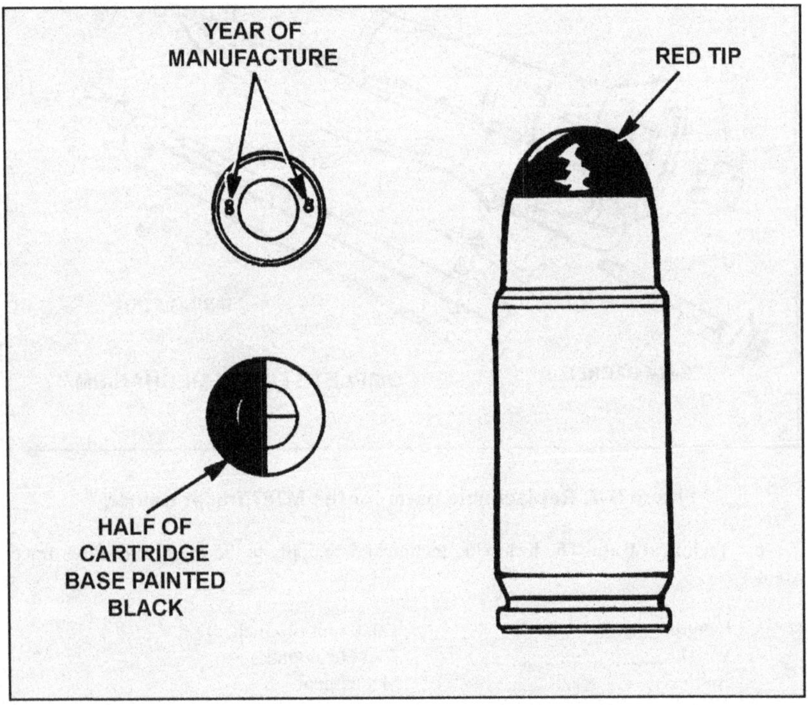

Figure B-8. M939, 9-mm tracer cartridge.

d. **Function Check**. Before the M287 is fired, a function check must be performed to ensure the trigger and safety mechanisms are operating properly. Function check procedures are shown in Table B-1. However, before performing a function check, ensure—

- The *cocking lever* is in the safe ("S") position.
- The *transport safety pin* is fully inserted, with the lanyard wrapped clockwise around the launcher.
- The *bolt* is removed from the breech.

STEP	OPERATOR ACTIONS	FUNCTION CHECK	CORRECTIVE MEASURES
1	Try to cock the tracer trainer	It should not cock	If it cocks, turn the trainer in for repair
2	Remove the transport safety pin and depress the forward safety	The forward safety should spring back when released	If it does not spring back, turn the tracer trainer in for repair
3	Cock the firing mechanism, pressing only the red trigger button	The firing rod should move only slightly and should be captured by the safety lever assembly. No more than half the rod should protrude through the rear of the firing assembly	If the tracer trainer fires, turn it in for repair
4	Recock the firing mechanism; fully depress and continue to hold down the forward safety; press the red trigger button	The tracer trainer should fire	If the tracer trainer *fails* to fire, turn it in for repair
5	Return the cocking lever to the SAFE position and reinstall the transport safety pin		

Table B-1. Function check, M287 tracer trainer.

WARNING
LOAD LIVE AMMUNITION ONLY ON THE FIRING LINE. NEVER FIRE THE M287 TRACER TRAINER AT HARD TARGETS LESS THAN 125 METERS FROM THE FIRING LINE. REMAIN CLEAR OF THE FRONT OF THE M287, WHICH MUST BE POINTED DOWNRANGE AT ALL TIMES.

e. **Loading**. The M287 tracer trainer is loaded as follows:

(1) The firer performs the function check and places the M287 tracer trainer on his shoulder.

(2) The trainer looks through the barrel from the rear to verify that it contains no obstructions. He inspects the cartridge primer to ensure it is not dented, inserts the bolt into the breech (Figure B-9), and presses in and turns the bolt clockwise to the "S" to

place it in the safe position (Figure B-10). Then he arms the tracer trainer by turning the bolt clockwise to "F" so that it can be fired (Figure B-11).

(3) With the launcher still on his shoulder, the firer cocks the launcher by placing the cocking lever in the cocked position.

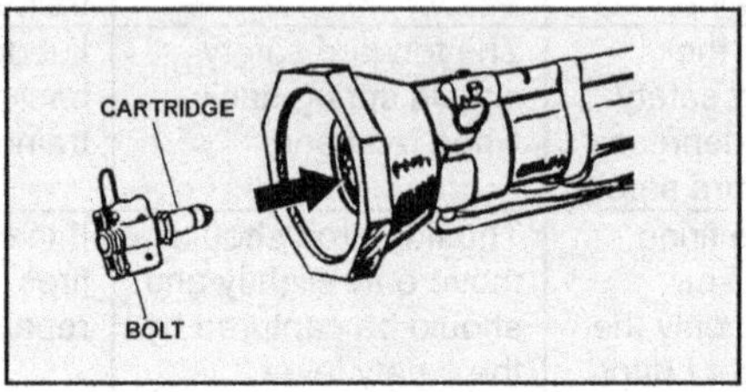

Figure B-9. Inserting bolt with cartridge into the breech of the M287 tracer trainer.

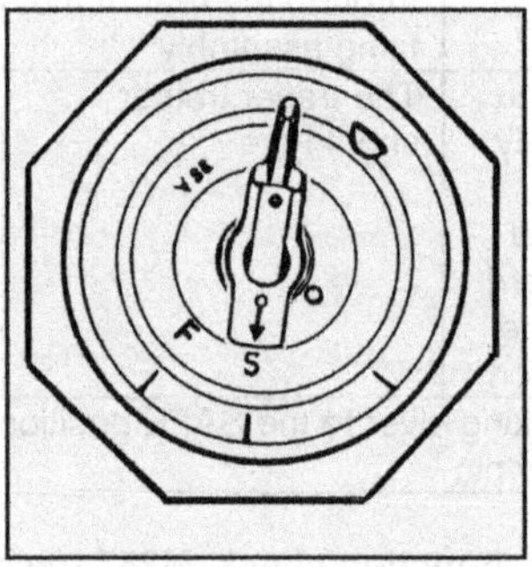

Figure B-10. Ensuring tracer trainer bolt is in the safe position.

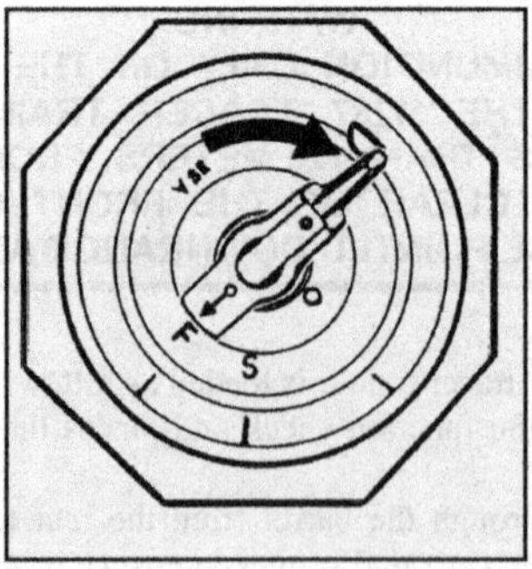

Figure B-11. Arming the tracer trainer.

f. **Unloading**. With the tracer trainer still on the firer's shoulder, the trainer or instructor removes the bolt by turning it counterclockwise past the "S," then pulling the bolt from the breech (Figure B-12). If more rounds are to be fired, he removes the expended cartridge and reloads the bolt.

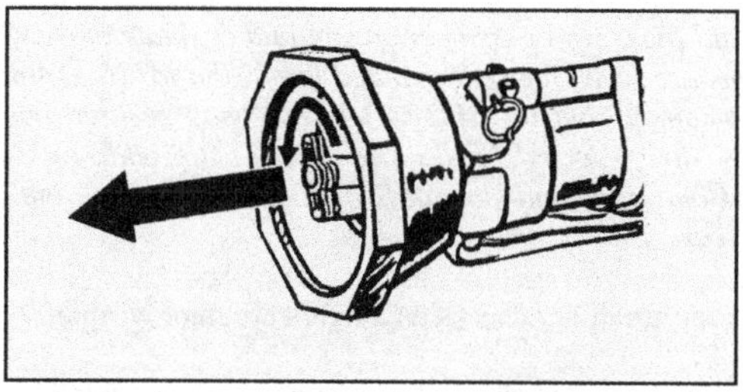

Figure B-12. Unloading the tracer trainer.

g. **Reloading**. With the tracer trainer still on his shoulder, the firer places the cocking lever in the safe position, keeping the tracer trainer on his shoulder and pointing toward the target. The trainer reloads the tracer trainer and prepares it for firing.

DANGER
NEVER TOUCH THE TRIGGER WHILE RELOADING.

h. **Misfire Procedures**. In the event of a misfire—

(1) Immediately shout "Misfire!"

(2) Release the forward safety, recock the launcher, and ensure the cocking lever is erect. Fully depress and hold down the forward safety, then aim and press the red trigger button.

(3) If the tracer trainer still fails to fire, repeat the two previous steps.

(4) If the launcher fails to fire again, release the forward safety, maintain your firing position, and return the cocking lever to the safe position.

(5) Notify the training supervisor.

(6) If a real misfire occurs that cannot be corrected by following the steps previously discussed, the trainer or assistant trainer on duty must ensure the cocking lever is in the safe position and that the forward safety is in the vertical position. He must then insert the transport safety pin and remove and inspect the cartridge. If the primer is dented, he replaces it, disposing of the old one IAW range safety SOP. If the primer is not dented, he notifies DS maintenance to inspect the bolt firing pin for damage. He then reloads the tracer trainer and continues the training exercise.

i. **Range Procedures and Safety**. These are the same as for the M136 AT4.

APPENDIX C

PERFORMANCE EVALUATIONS

The trainer administers performance evaluations to determine how much training the soldiers have learned. Those who fail are retrained and retested, and those who pass help retrain and evaluate those who did not. Instructions are given for right-handed firers: the M136 AT4 may only be fired right-handed, but the M72-series LAW may be fired left-handed, if procedures are reversed. Tasks are generic to both light antiarmor weapons. Conditions and standards may also be generic but, if not, are given by weapon.

TASK 1: Perform serviceability checks on a light antiarmor weapon.

CONDITIONS: Given an expended light antiarmor weapon under suitable conditions or in an appropriate location.

M72-Series LAW

STANDARDS: The firer performs the following actions:

TRAINING AND EVALUATION

1. Inspects the launcher to ensure it has no cracks, dents, or bulges.

2. Ensures the trigger arming handle is present and in the SAFE position.

3. Checks the rubber boots around the trigger bar and detent for cracks, tears, and deterioration.

4. (*M72A2 only*) Inspects the data plate on the right forward side of the launcher to ensure the words "with coupler" appear on the first line. If not, informs the trainer.

M136 AT4

STANDARDS: The firer performs the following actions:

TRAINING AND EVALUATION

1. Ensures the launcher has no cracks, dents, or bulges, or reports these defects to the instructor.

2. Ensures the transport safety pin is in place, or reports this defect to the instructor.

3. Ensures the cocking lever is in the SAFE (uncocked) position, or reports this defect to the instructor.

4. Ensures the muzzle cover is intact, or reports this defect to the instructor.

5. Ensures the color-coded band is correct, or reports this defect to the instructor.

6. Ensures the sights function properly, or reports this defect to the instructor.

7. Ensures the venturi has no foreign objects, or reports this defect to the instructor.

8. Ensures the forward safety functions properly, or reports this defect to the instructor.

9. Ensures the red trigger button is not broken or missing, or reports this defect to the instructor.

10. Ensures the shoulder stop is not broken or damaged, or reports this defect to the instructor.

11. Ensures the carrying sling is not frayed and is firmly attached to the launch tube, or reports this defect to the instructor.

RESULTS **GO** **NO-GO**

Critique and grade the soldier.

TASK 2: Place a light antiarmor weapon into operation.

CONDITIONS: Given an expended light antiarmor weapon (FHT) on a range or in garrison.

M72-Series LAW

STANDARDS: The firer performs the following actions, in sequence:

TRAINING AND EVALUATION

1. Removes the rear cover pull pin.

2. Releases the rear cover and sling assembly.

3. Extends the launcher vigorously and ensures it is fully extended.

4. Places the launcher on his firing shoulder.

5. Checks the backblast area.

6. Pulls the arming handle out.

7. Sights.

M136 AT4

STANDARDS: The firer performs the following actions, in sequence:

TRAINING AND EVALUATION

1. Cradles the AT4 in his arms.

2. Removes the transport safety pin.

3. Unsnaps and unfolds the shoulder stop.

4. Places the AT4 on his shoulder.

5. Releases the sights.

6. Checks the backblast area.

7. Places cocking lever in the ARMED (cocked) position.

8. Fully depresses and holds down the forward safety.

RESULTS **GO** **NO-GO**

Critique and grade the soldier.

TASK 3: Demonstrate the standing position used to fire a light antiarmor weapon.

CONDITIONS: Given a target and an expended light antiarmor weapon in any suitable location.

STANDARDS: The firer demonstrates the standing position as follows:

TRAINING AND EVALUATION

M72-Series LAW

1. Faces the target by executing a half-left face turn and spreading feet a comfortable distance apart.

2. Places the launcher on the shoulder with the left hand directly under the forward portion of the launcher.

3. Places the right elbow against the body for stability.

M136 AT4

1. Faces the target by executing a half-left face turn and spreads the feet a comfortable distance apart.

2. Places the launcher on the shoulder and grasps the sling near the launcher's muzzle with the left hand.

3. Places the right elbow against the body for stability.

RESULTS **GO** **NO-GO**

Critique and grade the soldier.

TASK 4: Demonstrate the kneeling position used to fire a light antiarmor weapon.

CONDITIONS: Given a target and an expended light antiarmor weapon in a suitable location, with the soldier in the standing position.

STANDARDS: The firer demonstrates the kneeling position as follows:

TRAINING AND EVALUATION

1. Kneels on the right knee with the right thigh at a 90-degree angle to the line of aim.

2. Sits back on the right heel and shifts weight forward.

3. Rests the upper left arm forward of the left knee.

4. Places the right arm against the body.

RESULTS **GO** **NO-GO**

Critique and grade the soldier.

TASK 5: Demonstrate the sitting position used to fire a light antiarmor weapon.

CONDITIONS: Given a light antiarmor weapon training device or expended launcher in a suitable location, with the soldier in the standing position.

STANDARDS: Each firer demonstrates the sitting position as follows:

TRAINING AND EVALUATION

1. Keeps the launcher pointed toward the target.

2. Executes a half-left face turn and sits with the legs crossed.

3. Places the launcher on the right shoulder.

4. Leans slightly forward from the hips.

5. Rests the elbows forward of the knees to avoid bone-to-bone contact.

RESULTS **GO** **NO-GO**

Critique and grade the soldier.

TASK 6: Demonstrate the prone position used to fire a light antiarmor weapon.

CONDITIONS: Given a target and an expended light antiarmor weapon (FHT) in a suitable location.

STANDARDS: The firer demonstrates the prone position as follows:

TRAINING AND EVALUATION

1. To keep clear of the backblast area, lies down at an angle of not less than 45 degrees from the line of fire.

2. Keeps back straight and right leg directly on line with the right hip and right shoulder.

3. Moves left leg as far as possible to ensure comfort.

4. Holds both elbows well below the launcher.

5. Checks the backblast area to ensure that no part of the body is in it.

6. Holds head as steady as possible, with the firing eye lined up with the sights.

RESULTS	GO	NO-GO

Critique and grade the soldier.

TASK 7: Identify enemy vehicles and weapons.

CONDITIONS: Given ten vehicle identification cards (GTA 17-2-13).

STANDARDS: The soldier correctly identifies eight out of ten vehicles or weapons by nomenclature.

RESULTS	GO	NO-GO

Critique and grade the soldier.

TASK 8: Estimate range.

CONDITIONS: Given stationary, partly exposed, or fully exposed targets at ranges between 50 and 250 meters, during daylight, in weather conditions where all targets are visible. Targets include personnel, equipment, silhouettes, and vehicles.

STANDARDS: The soldier states the distance to each target with no more than 20 percent error.

RESULTS	GO	NO-GO

Critique and grade the soldier.

TASK 9: Apply the correct lead for a target.

CONDITIONS: Given a sight template, a series of ten flank, frontal, and oblique silhouettes of armored targets, rates of movement, and ranges. (At least three different views, speeds, and ranges are given.)

STANDARDS: The soldier uses correct sighting procedures and correctly applies the lead eight out of ten times.

RESULTS	GO	NO-GO

Critique and grade the soldier.

TASK 10: Perform misfire procedures on a light antiarmor weapon.

CONDITIONS: Given a previously fired light antiarmor weapon and instructions that the weapon has misfired, on a suitable firing range, under simulated combat conditions, in any lighting conditions.

M72-series LAW

STANDARDS: The soldier applies misfire procedures as follows and tries to fire the M72-series LAW. If the weapon still fails to fire, he disposes of it IAW unit SOP:

TRAINING AND EVALUATION

1. Immediately squeezes the trigger bar again.

2. If the LAW still does not fire, tries to place the trigger arming handle on SAFE.

3. Removes the launcher from the shoulder and partly collapses it (10 to 15 centimeters or 4 to 6 inches), keeping the launcher pointed upward and downrange.

4. Extends the launcher.

5. If the LAW fails to fire again, squeezes the trigger, tries to return the weapon to SAFE, partly collapses the launcher, and sets the weapon aside.

M136 AT4

STANDARDS: The soldier performs combat misfire procedures on the AT4 as follows:

TRAINING AND EVALUATION

1. Releases the forward safety with the right hand.

2. Recocks the firing mechanism using the cocking lever.

3. Presses and holds down the forward safety, aims, and presses the trigger.

4. If launcher still fails to fire, repeats Steps 1 and 2.

5. If the launcher again fails to fire, returns the cocking lever to the SAFE (uncocked) position.

6. Removes the weapon from the shoulder, keeping it pointed toward the target.

7. Replaces the transport safety pin.

8. Informs the instructor that the firer breaks the sights off to indicate a misfired weapon when in combat.

9. Places the launcher on the ground, pointing toward the target.

RESULTS	GO	NO-GO

Critique and grade the soldier.

TASK 11: Return the light antiarmor weapon to its carrying configuration.

CONDITIONS: Given an expended light antiarmor weapon in a suitable location.

STANDARDS: The soldier performs the following actions, in sequence:

M72-series LAW

TRAINING AND EVALUATION

1. Pushes in the trigger arming handle.

2. Removes the launcher from the shoulder.

3. Depresses the barrel detent and collapses the launch tube, guiding the front and rear sights into position.

4. Replaces the sling assembly.

5. Replaces the rear cover pull pin.

M136 AT4

1. Releases the forward safety.

2. Returns the cocking lever to the SAFE (uncocked) position by pushing forward and upward on it, then letting it spring back into position.

3. Removes the launcher from the shoulder, ensuring the muzzle is pointed in the direction of fire.

4. Replaces the transport safety pin, fully seating it in the retainer hole.

5. When closing the rear sight, resets the range indicator to the 200-meter setting. (If this step is not accomplished, the rear sight breaks off when the sight cover is closed.)

6. Restores the sights by laying them down and closing their covers.

7. Snaps the shoulder stop into the closed position.

8. Moves to another location; slings the launcher over either shoulder and moves out.

RESULTS **GO** **NO-GO**

Critique and grade the soldier.

APPENDIX D

SUGGESTED DESIGN FOR COMBINED LIGHT ANTIARMOR RANGE

This appendix provides basic guidance for units to establish a range complex for light antiarmor weapons. The complex described may be used for multilevel training and firing of light antiarmor weapons; both subcaliber trainers and live HE ammunition may be used on the same range. Training areas should be near, but not adjacent to, the firing line. The firing line should be designed to allow personnel to observe firing from the side. The target array should include stationary targets composed of F-type targets or the hulls of hard targets (APCs or tanks), a moving target (a target on track or a manned moving target tank [MMTT]), and bunkers. This allows the firer to engage a variety of targets and to develop his skills in the various types of engagements.

D-1. FIRING LINE

The firing line should be organized to allow for the 100-meter AT4 backblast (Figure D-1). Also, at least 20 meters must be allowed between firing points. The backblast area must be fenced, roped, or marked in some way, so soldiers know not to enter it when firing is being conducted. A range that meets the requirements of the AT4 will also meet the requirements of the LAW.

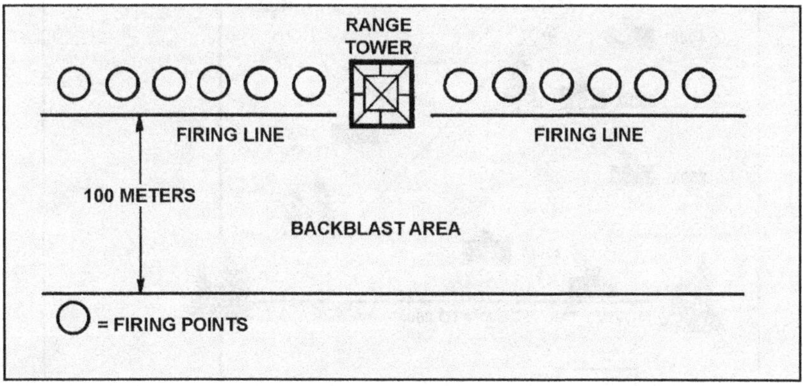

Figure D-1. Firing line.

D-2. TARGET ARRAY

The target array should include targets at ranges of 100 to 700 meters (Figure D-2). This enables the unit to conduct advanced instructional fire on targets located beyond the weapons' maximum effective ranges. Standard vehicle silhouettes or tank hulls should be used for stationary targets. Selected targets should be positioned behind mounds of earth to simulate vehicles in defilade or partly behind trees or other objects. Firers adjust their points of aim accordingly. Targets should also be positioned to allow firers to engage flank, frontal, and oblique targets. Bunker targets should be placed within the target array at ranges of 100 to 400 meters. Bunker targets should be no more than 1 meter high or 2 1/2 to 3 meters wide. Moving target silhouettes should travel along a track or road so the firers can engage fast moving targets from both flank and oblique angles. The moving targets should be placed at ranges of 150 to 600 meters and should enter and exit from the flanks outside the range fan. Firers may only use HE weapons to engage hard targets. This reduces the damage to other targets the unit must maintain such as F-type silhouettes and bunkers.

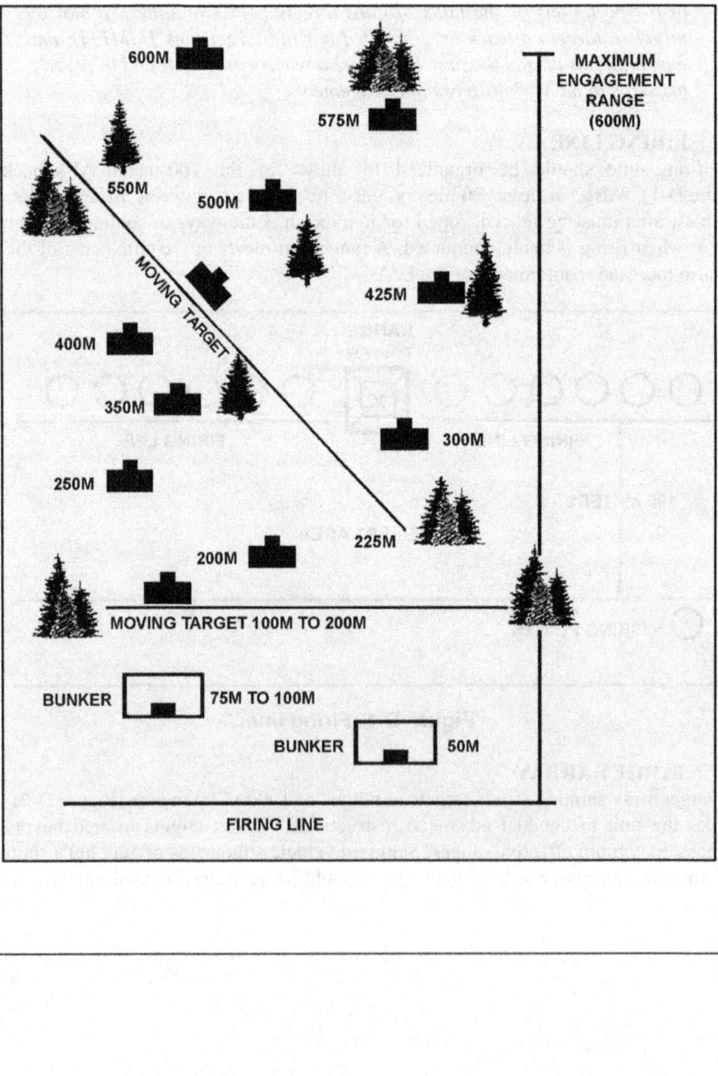

Figure D-2. Target array.

APPENDIX E
FIRING TABLES AND EXAMPLE SCORECARDS

This appendix provides firing tables for the subcaliber gunnery trainers discussed in this manual. It also includes an example of a completed scorecard for each light antiarmor weapon. All instructional and qualification firing tables are conducted on suitable ranges with ammunition designated in the firing tables. Some tables are fired in daylight, others in limited visibility, but all have some NBC firing (Table E-1).

M72-Series LAW	Instructional day fire, stationary targets
	Instructional day fire, moving targets
	Instructional night fire
M136 AT4	Instructional fire
	Instructional night fire
	Qualification fire
	Qualification night fire
	Advanced instructional fire

Table E-1. Firing tables for light antiarmor weapons.

E-1. M72-SERIES LAW

The M190 35-mm subcaliber device is used to fire the instructional firing tables for the M72-series LAW and DA Form 7323-R (M72-Series LAW Scorecard) is used to score them. Figure E-1 shows an example of the completed form. The instructional day and night firing tables for the LAW are provided in Tables E-2 and E-3.

NOTE: A blank copy of the form is located at the end of this publication for local reproduction on 8 1/2-inch by 11-inch paper.

M72-SERIES LAW SCORECARD

Instructions for using this form are provided in FM 23-25.
TRADOC is the proponent agency.

NAME *THIELE, STEPHEN* SSN *123-45-6789* DATE *19 APR 94*

GRADE *E-5* RATING *1ST CLASS* UNIT *HHC 2/29 INF. REGT.* WEATHER (circle one) CLEAR **(RAIN)** FOG SNOW

DATA REQUIRED BY THE PRIVACY ACT OF 1974

AUTHORITY: 10 USC 3012(g) / Executive Order 9397 **PRINCIPAL PURPOSE(S):** Record individual performance. **ROUTINE USES:** Evaluate individual proficiency and determine proficiency level. SSN is used for positive identification purposes only. **MANDATORY OR VOLUNTARY DISCLOSURE AND EFFECT ON INDIVIDUAL NOT PROVIDING INFORMATION:** Voluntary. Individuals not providing information cannot be rated/scored on a mass basis.

TABLE	TYPE OF FIRE	TYPE OF TARGET	ROUND	RANGE (M)	FIRING POSITION	HIT	MISS
1	Instructional DAY	Stationary	1	125 to 150	Standing	✓	
			2	150 to 175	Kneeling	✓	
			* 3	150 to 200	Sitting	✓	
			* 4	125 to 150	Prone	✓	
					Table Subtotals	4	0

FIRER'S SIGNATURE [signature] DATE *19 APR 94* SCORER'S SIGNATURE [signature] DATE *19 APR 94*

TABLE	TYPE OF FIRE	TYPE OF TARGET	ROUND	RANGE (M)	FIRING POSITION	HIT	MISS
2	Instructional DAY	Moving	1	100 to 200	Standing	✓	
			2	100 to 200	Sitting	✓	
			* 3	100 to 200	Kneeling		✓
					Table Subtotals	2	1

FIRER'S SIGNATURE [signature] DATE *20 APR 94* SCORER'S SIGNATURE [signature] DATE *20 APR 94*

TABLE	TYPE OF FIRE	TYPE OF TARGET	ROUND	RANGE (M)	FIRING POSITION	HIT	MISS
3	Instructional NIGHT	Stationary	1	100 to 150	Kneeling		✓
			2	100 to 150	Sitting	✓	
			* 3	100 to 150	Sitting	✓	
					Table Subtotals	2	1

FIRER'S SIGNATURE [signature] DATE *21 APR 94* SCORER'S SIGNATURE [signature] DATE *21 APR 94*

* Firer must wear a protective mask when firing this round.

CONSOLIDATED SCORES

FIRER'S SIGNATURE [signature] DATE *21 APR 94* TOTAL MISSES *2*

SCORER'S SIGNATURE [signature] DATE *21 APR 94* RANK *SFC*

RANGE OIC'S SIGNATURE [signature] DATE *21 APR 94* RANK *CPT*

INSTRUCTOR NOTES

ALL FIRING TABLES
- Do not announce the ranges to targets.
- Fire all engagements in sequence.

DA Form 7323, JUL 94

Figure E-1. Example completed DA Form 7323-R.

TABLE 1 M72-Series LAW Instructional Day Fire, Stationary Targets	
TASK	Each soldier engages stationary armored targets with the M190 subcaliber device.
CONDITIONS	Given one M190 subcaliber device and four M73 rockets, on a suitable firing range with stationary targets at a range of 100 to 225 meters. Two targets are engaged in MOPP4.
STANDARD	The soldier fires four rockets at stationary targets and achieves at least three hits.
TABLE 2 M72-Series LAW Instructional Day Fire, Moving Targets	
TASK	Each soldier engages moving armored targets with the M190 subcaliber device.
CONDITIONS	Given one M190 subcaliber device, three M73 rockets, and targets moving at a rate of 8 to 24 kmph and at a range of 100 to 200 meters. Frontal and oblique-moving targets may be used when a manned, moving tank is available. One target is engaged in MOPP4.
STANDARD	The soldier fires three rockets at moving targets and achieves at least two hits.

Table E-2. Instructional day firing tables for the M72-series LAW.

TABLE 3 M72-Series LAW Instructional Night Firing Before you can use the AN/PVS-4, you must install an M72A1 sight reticle on it and Boresight it to an expended M72-series LAW launcher.	
TASK	Each soldier engages stationary targets at night with the M190 subcaliber device.
CONDITIONS	Given one M190 subcaliber device and three M73 rockets, on a suitable firing range with stationary targets at a range of 100 to 150 meters, at night, and either an AN/PVS-4 or illumination provided by indirect fire.
STANDARD	The soldier fires three rockets at stationary targets and achieves at least one hit. He engages one of these targets in MOPP4.

Table E-3. Instructional night firing table for the M72-series LAW.

E-2. M136 AT4

The M287 9-mm tracer bullet trainer is used to fire the instructional, qualification, and advanced instructional firing tables for the M136 AT4 and DA Form 7324-R (M136 AT4 Scorecard) is used to score them. A copy of this form is located at the end of this publication for local reproduction on 8 1/2-inch by 11-inch paper. Figure E-2 shows an example of the completed form; Tables E-4 through E-6, on pages E-7, E-8, and E-9, provide the instructional, qualification, and advanced instructional firing tables for the AT4. The latter must be fired on a separate, more specialized range.

E-3. NIGHT COURSE OF FIRE TABLES AND QUALIFICATION FOR M136 AT4

Each soldier is given a PMI before the instructional and qualification tables are fired. Once the soldier completes Table I and III (day tables), he is given additional instruction on the use of the aided vision devices for the night operations.

a. **Qualification Tables for the M287 Tracer Trainer.**

- Table I (Instructional Day). This scenario consists of 12 targets from 100 to 300 meters. The standard to proceed to the next table is 6 out of 12 targets successfully engaged.
- Table II (Instructional Night). This scenario consists of 8 targets from 100 to 300 meters. The standard to proceed to the next table is 6 out of 12 targets successfully engaged at night.
- Table III (Qualification Day). This scenario consists of 8 targets from 100 to 300 meters. The standard to proceed to the next table is 6 out of 8 targets successfully engaged.
- Table IV (Qualification Night). This scenario consists of 8 targets from 100 to 300 meters (same as Table III, but conducted at night). The standard to proceed to the advanced firing tables is 6 out of 8 targets successfully engaged.

NOTES: 1. In Table IV, eight engagements are proposed as opposed to the current six engagements.
2. All references to the M136 launcher and cartridge refer to the M287 9-mm tracer bullet training device.

b. **Ammunition Requirements for Night Training Strategy**. The current STRAC ammunition allocation and course of fire is compared with the proposed night training strategy in Table E-7 on page E-9. The current STRAC table stipulates that 120 rounds of 9-mm tracer ammunition is authorized to each squad for additional squad members to fire Tables I and IV.

c. **Day Standards**. Commanders are able to achieve the current day standard at night with the M136 AT4 through the use of training strategy within this field manual. A gunner who is qualified on the day standard and has executed preliminary marksmanship training with the appropriate night vision device can achieve the current day standard at night. During testing, gunners had no problem engaging targets out to 300 meters. Table E-8, on page E-9, shows the day standard and is also the proposed night standard.

M136 AT4 SCORECARD

Instructions for using this form are provided in FM 23-25.
TRADOC is the proponent agency.

NAME DINKENS, DAVID SSN 135-24-5678 DATE 19 APR 94

GRADE E-3 RATING ______ UNIT HHC 2/29 INF REGT. WEATHER (circle one) CLEAR (RAIN) FOG SNOW

DATA REQUIRED BY THE PRIVACY ACT OF 1974

AUTHORITY: 10 USC 3012(g) / Executive Order 9397. **PRINCIPAL PURPOSE(S):** Record individual performance. **ROUTINE USES:** Evaluate individual proficiency and determine proficiency level. SSN is used for positive identification purposes only. **MANDATORY OR VOLUNTARY DISCLOSURE AND EFFECT ON INDIVIDUAL NOT PROVIDING INFORMATION:** Voluntary. Individuals not providing information cannot be rated/scored on a mass basis.

TABLE	TYPE OF FIRE	TYPE OF TARGET	ROUND	RANGE (M)	FIRING POSITION	HIT	MISS
1	Instructional Day	Stationary	1	150 to 200	Kneeling	✓	
		Stationary	* 2	150 to 200	Kneeling	✓	
		Stationary	3	250 to 300	Sitting	✓	
		Stationary	4	250 to 300	Sitting	✓	
		Stationary	5	125 to 150	Prone	✓	
		Stationary	6	125 to 150	Prone	✓	
		MMTT/MAS	7	150 to 200	Kneeling	✓	
		MMTT/MAS	* 8	150 to 200	Kneeling	✓	
		MMTT/MAS	9	200 to 250	Standing		✓
		MMTT/MAS	*10	200 to 250	Fighting position, unsupported		✓
		MMTT/MAS	11	250 to 300	Kneeling	✓	
		MMTT/MAS	12	250 to 300	Kneeling		✓
					Table Subtotals	9	3

FIRER'S SIGNATURE David Dinkens DATE 19 APR 94 SCORER'S SIGNATURE Dan Sell DATE 19 APR 94

TABLE	TYPE OF FIRE	TYPE OF TARGET	ROUND	RANGE (M)	FIRING POSITION	HIT	MISS
2	Instructional Night	Stationary	1	125 to 150	Prone		✓
		Stationary	2	125 to 150	Prone	✓	
		Stationary	* 3	150 to 200	Sitting	✓	
		Stationary	4	150 to 200	Sitting	✓	
		Stationary	5	250 to 300	Kneeling		✓
		Stationary	6	250 to 300	Kneeling	✓	
		MMTT/MAS	7	125 to 150	Standing	✓	
		MMTT/MAS	8	125 to 150	Standing	✓	
		MMTT/MAS	* 9	200 to 250	Kneeling	✓	
		MMTT/MAS	10	200 to 250	Kneeling	✓	
		MMTT/MAS	11	150 to 200	Standing		✓
		MMTT/MAS	12	150 to 200	Fighting position, unsupported	✓	
					Table Subtotals	9	3

FIRER'S SIGNATURE David Dinkens DATE 19 APR 94 SCORER'S SIGNATURE Dan Sell DATE 19 APR 94

* Firer must wear a protective mask when firing this round.

INSTRUCTOR NOTES	
ALL FIRING TABLES	• Do not announce the ranges to targets. • Fire all engagements in sequence. • Fire Table 1 before Table 3; fire Table 2 before Table 4.
FIRING TABLES 2 THROUGH 5	• Ensure no one fires at front or rear of a manned moving tank target. • Total hits and misses, and enter results in the consolidated scores block on the back of this scorecard.

DA Form 7324-R, JUL 94 (FRONT)

Figure E-2. Example completed DA Form 7324-R (front).

M136 AT4 SCORECARD

TABLE	TYPE OF FIRE	TYPE OF TARGET	ROUND	RANGE (M)	FIRING POSITION	HIT	MISS
		Stationary	1	125 to 150	Standing	✓	—
		Stationary	2	150 to 200	Kneeling	✓	—
		Stationary	* 3	200 to 250	Sitting	✓	—
3	Qualification	Stationary	4	250 to 300	Sitting	✓	—
	Day	MMTT/MAS	5	125 to 150	Kneeling	✓	—
		MMTT/MAS	6	150 to 200	Fighting position, unsupported	—	✓
		MMTT/MAS	* 7	200 to 250	Sitting	✓	—
		MMTT/MAS	8	250 to 300	Kneeling	✓	—
					Table Subtotals	7	1

FIRER'S SIGNATURE *David Dinkens* DATE *21 Apr 94* SCORER'S SIGNATURE *Dan Sell* DATE *21 Apr 94*

TABLE	TYPE OF FIRE	TYPE OF TARGET	ROUND	RANGE (M)	FIRING POSITION	HIT	MISS
		Stationary	* 1	125 to 150	Standing or Prone	✓	—
		Stationary	2	200 to 250	Kneeling	✓	—
4	Qualification	Stationary	3	250 to 300	Sitting	✓	—
	Night	MMTT/MAS	4	150 to 200	Sitting	✓	—
		MMTT/MAS	* 5	125 to 150	Kneeling	—	✓
		MMTT/MAS	6	200 to 250	Fighting position, unsupported	✓	—
					Table Subtotals	5	1

FIRER'S SIGNATURE *David Dinkens* DATE *21 Apr 94* SCORER'S SIGNATURE *Dan Sell* DATE *21 Apr 94*

TABLE	TYPE OF FIRE	TYPE OF TARGET	ROUND	RANGE (M)	FIRING POSITION	HIT	MISS
		Sequence	1		Sitting	✓	—
		Sequence	2		Sitting	✓	—
5	Advanced	Pair	3		Kneeling	✓	—
	Instructional	Pair	4		Kneeling	✓	—
		Volley	5		Kneeling	✓	—
		Volley	* 6		Sitting	✓	—
					Table Subtotals	6	0

FIRER'S SIGNATURE *David Dinkens* DATE *22 Apr 94* SCORER'S SIGNATURE *Dan Sell* DATE *22 Apr 94*

* Firer must wear a protective mask when firing this round.

QUALIFICATION CRITERIA — BASED ON TOTAL HITS FOR TABLES 3 AND 4

FIRST CLASS .... 13 HITS
SECOND CLASS .... 11 HITS (circled)
THIRD CLASS .... 9 HITS
UNQUALIFIED .... FEWER THAN 9 HITS

CONSOLIDATED SCORES

INSTRUCTIONAL TOTALS (TABLES 1 AND 2): HITS *18* MISSES *6* CLASS RATING, TABLES 1 AND 2: *FIRST*

QUALIFICATION TOTALS (TABLES 3 AND 4): HITS *12* MISSES *2* OVERALL TOTALS HITS *30* MISSES *8*

FIRER'S SIGNATURE *David Dinkens* DATE *22 Apr 94* SCORER'S SIGNATURE *Dan Sell* DATE *22 Apr 94*

RANGE OIC'S SIGNATURE *[signature]* RANK *SFC* DATE *22 Apr 94*

DA Form 7324-R, JUL 94 (BACK)

Figure E-2. Example completed DA Form 7324-R (back).

TABLE 1 M136 AT4 Instructional Fire	
TASK	Each soldier engages an armored target with the M287 tracer bullet trainer.
CONDITIONS	Given one M287 9-mm subcaliber device and twelve rounds of 9-mm tracer ammunition, a suitable firing range, stationary targets at a range of 100 to 300 meters, and other targets moving 8 to 24 kmph at a range of 150 to 200 meters. When a manned, moving tank is available, frontal and oblique-moving targets may be used.
STANDARD	The soldier fires twelve tracer bullets at stationary and moving targets. He engages three of these targets in MOPP4.
TABLE 3 M136 AT4 Instructional Fire	
TASK	Each soldier engages an armored target with the M287 tracer bullet trainer at night.
CONDITIONS	At night, given one M287 9-mm subcaliber device and twelve rounds of 9-mm tracer ammunition, artificial illumination, a suitable firing range, stationary targets at a range of 100 to 300 meters, and other targets moving from 8 to 24 kmph (5 to 15 mph) at a range of 150 to 200 meters. When a manned, moving tank is available, frontal and oblique-moving targets may be used.
STANDARD	The soldier fires twelve tracer bullets at stationary and moving targets. He engages two of these targets in MOPP4.

Table E-4. Instructional firing tables for the M136 AT4.

TABLE 2 M136 AT4 Qualification Fire	
TASK	Each soldier engages an armored target with the M287 tracer bullet trainer.
CONDITIONS	Given one M287 9-mm subcaliber device and eight rounds of 9-mm tracer ammunition, a suitable firing range, stationary targets at a range of 100 to 300 meters, and other targets moving at a rate of 8 to 24 kmph (5 to 15 mph) at a range of 150 to 200 meters. When a manned, moving tank is available, frontal and oblique-moving targets may be used.
STANDARD	The soldier fires eight tracer bullets at stationary and moving targets and achieves at least six hits. He engages two of these targets in MOPP4.
TABLE 4 M136 AT4 Qualification Fire	
TASK	Each soldier engages an armored target with the M287 tracer bullet trainer.
CONDITIONS	At night, given one M287 9-mm subcaliber device and six rounds of 9-mm tracer ammunition, artificial illumination, a suitable firing range, stationary targets at a range of 100 to 300 meters, and other targets moving 8 to 24 kmph (5 to 15 mph) at a range of 150 to 200 meters. When a manned, moving tank is available, frontal and oblique-moving targets may be used.
STANDARD	At night, given one M287 9-mm subcaliber device and six rounds of 9-mm tracer ammunition, artificial illumination, a suitable firing range, stationary targets at a range of 100 to 300 meters, and other targets moving 8 to 24 kmph (5 to 15 mph) at a range of 150 to 200 meters. When a manned, moving tank is available, frontal and oblique-moving targets may be used.

Table E-5. Qualification firing tables for the M136 AT4.

TABLE 5 M136 AT4 Advanced Instructional Fire Fire this table on an appropriate separate, specialized range.	
TASK	Each soldier engages an armored target with the M287 tracer bullet trainer.
CONDITIONS	Given one M287 9-mm subcaliber device and six rounds of 9-mm tracer ammunition, a suitable firing range, stationary targets at a range of 100 to 300 meters, and other targets moving 8 to 24 kmph (5 to 15 mph) at a range of 150 to 200 meters. When a manned, moving tank is available, frontal and oblique-moving targets may be used.
STANDARD	The soldier must fire six tracer bullets at stationary and moving targets and achieve at least three hits. He must engage two of the targets with sequence firing, two with pair firing, and two with volley firing, and he must observe MOPP4 while engaging one of the six.

Table E-6. Advanced instructional firing table for the M136 AT4.

EVENT	CURRENT STRAC	RECOMMENDED CHANGES	TOTAL
TABLE I	12	0	12
TABLE II	12	0	12
TABLE III (QUALIFICATION)	8	0	8
TABLE IV (NIGHT QUALIFICATION)	6	2	8
TABLE V (ADVANCE)	6	0	0
TOTAL	44	2	40

Table E-7. Ammunition requirements for night training strategy.

M136 AT4 (M287 TRAINER) TABLE III/IV			
ENGAGEMENT	**RANGE**	**POSITION**	**HIT/MISS**
1	125 to 150	Standing	
2	150 to 200	Kneeling	
3	200 to 250	Sitting	
4	250 to 300	Sitting	
5	125 to 150	Kneeling	
6	150 to 200	Fighting Position	
7	200 to 250	Sitting	
8	250 to 300	Kneeling	

Table E-8. M136 AT4 (M287 trainer) Table III/IV.

APPENDIX F
INFRARED LASER AIMING DEVICES

The newest infrared laser pointers increase the night firing accuracy of infantry weapons. The new family of weapon mounts (some fielded with the thermal weapon sight) and helmet brackets are more stable and reliable for continued accurate firing under field conditions. Their effectiveness is limited by the capability of the image-intensifying (I^2) sight with which they are used. Normally this is a night vision goggle or monocular, with or without the 3X magnifier. Again, they primarily enhance the gunner's ability to align his sights on the target, but usually have dramatic affect on his firing positions and scanning technique. Unlike I^2 sights mounted on the weapon, the gunner can somewhat reduce the adverse effects of muzzle flash by rising up higher off the sight with his helmet-mounted goggle or monocular. With the AN/PEQ-2A, the range of these devices is greatly improved, allowing great effectiveness for crew-served weapons out to 2,200 meters. However, trainers and leaders must adopt new safety procedures to ensure these lasers, which are not safe for the eyes, are treated like loaded weapons any time the batteries are issued. Finally, the bore light is also effective to boresight these pointers. Generally weapons do not require zeroing with live ammunition to confirm alignment after soldiers and leaders are qualified in the boresighting techniques. The AT4 cannot be conventionally boresighed, therefore a sight alignment method is used to zero the aided vision device.

F-1. AN/PAQ-4C, AIMING LIGHT

The AN/PAQ-4C aiming light (Figure F-1) projects an infrared laser beam, which cannot be seen with the eye but can be seen with night vision devices. This aiming light works with the AN/PVS-7B/C/D-series goggles and AN/PVS-14. The AN/PAQ-4C mounts on various weapons with mounting brackets and adapters.

a. **Technical Data**. The following technical data apply to the AN/PAQ-4C:

Optics:	100% parallax free, anti-reflective coated lens system
Length (sight):	14 centimeters (5.5 inches)
Weight:	164 grams (5.78 ounces)
Height:	3 centimeters (1.2 inches)
Width:	6.5 centimeters (2.5 inches)
Ranged:	Beyond 600 meters. (Actual range depends on light level and night vision device used for observation.)
Battery life:	100-hour operating (ON) time for AA batteries in temperatures above 0 degrees Centigrade (32 degrees Fahrenheit); 36 hours for temperatures below 0 degrees Centigrade (32 degrees Fahrenheit)

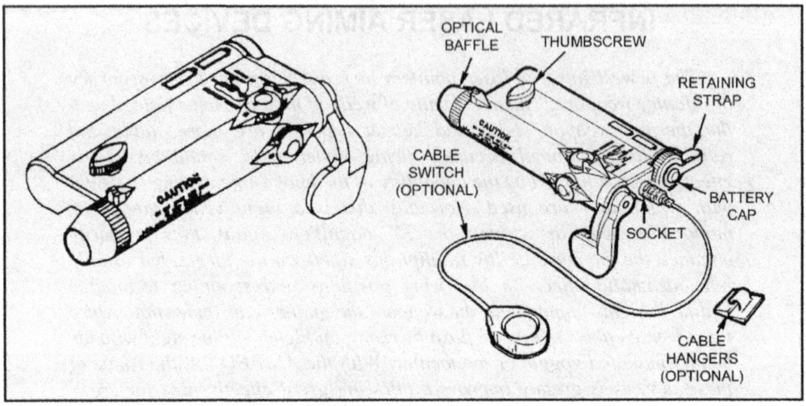

Figure F-1. AN/PAQ-4C, aiming light.

b. **Operation.** Activate the AN/PAQ-4C aiming light by rotating the ON/OFF switch lever or the button on the optional cable switch. Either switch connects power from two AA batteries to an internal electronic circuit, which produces the infrared laser. Internal lenses focus the infrared light into a narrow beam. Control the direction of the beam by rotating the mechanical adjusters with click detents. These adjusters are used to zero the aiming light to the weapon. Once zeroed to the weapon, the aiming light projects the beam along the line of fire of the weapon. The optical baffle prevents off-axis viewing of the aiming light beam by the enemy.

(1) ***M136 (AT4) Launcher and Cartridge Mounting Procedures***. The aiming light is attached to the M136 (AT4) launcher and cartridge by first attaching the mounting bracket to the weapon and then attaching the aiming light to the mounting bracket (Figures F-2 and F3).

STEP 1: Install the mounting bracket.

WARNING

Only M136 (AT4) qualified personnel should install and use the mounting bracket assembly. Conduct all M136 (AT4) operator preventive maintenance checks and services before installation of mounting bracket assembly.

NOTE: All references to the M136 (AT4) launcher and cartridge are also applicable to the M287 9-mm tracer bullet training device.

(a) Cradle the M136 (AT4) in left arm.

(b) Position the support bracket with the mounting rail on the left side and the marking FRONT over the rear sight.

(c) With the pivot bracket spread open, place the support bracket against the base of the rear sight housing and the bottom on the shoulder strap boss.

(d) Swing the pivot bracket around the M136 (AT4) and secure it by rotating the locking latch clockwise to engage the latch shaft.

(e) The lever screw assembly must be located in the rear threaded screw hole when mounting the aiming light. Lever screw assembly may require relocation from the front to the rear threaded screw hole.

(f) Place the bracket adapter (Figure F-2) in groove of the mounting rail so that the threaded screw hole in the base of the adapter is aligned with the lever screw assembly and tighten the lever screw.

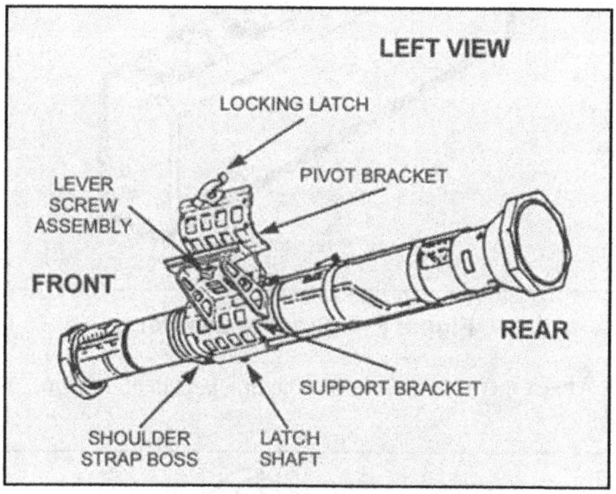

Figure F-2. M136 (AT4) installation.

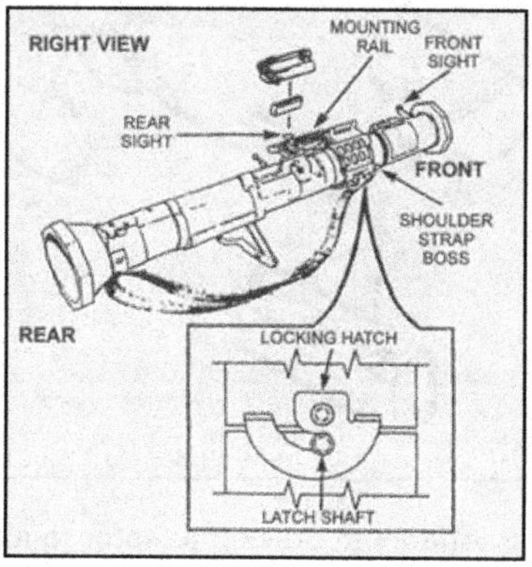

Figure F-3. M136 (AT4) installation (locking latch and shaft).

STEP 2: Install the bracket adapter (Figure F-4). The bracket adapter provides compatibility between the aiming light and the mounting brackets used on the M2, M60, and M136 AT4 weapons.

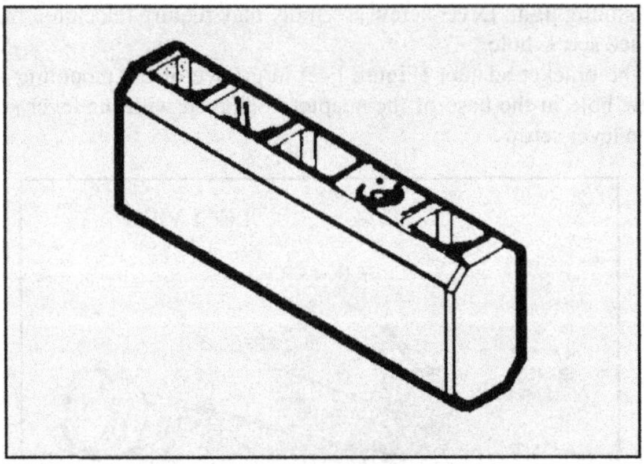

Figure F-4. Bracket adapter.

STEP 3: Attach the AN/PAQ-4 onto the bracket adapter (Figure F-5).

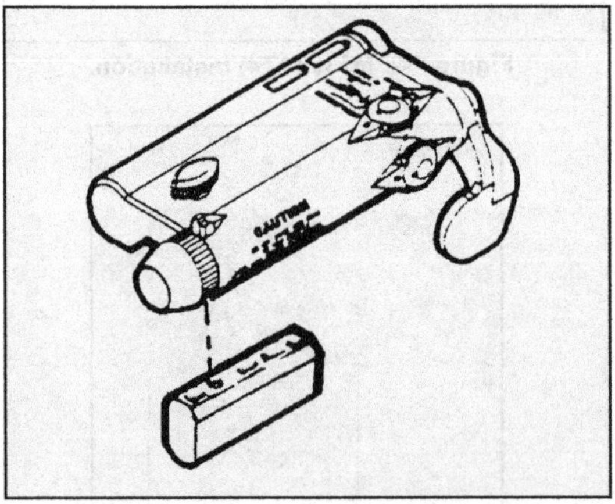

Figure F-5. Attaching bracket adapter to aiming light.

STEP 4: Attach the AN/PAQ-4 to the AN/PVS-4 mounting bracket.

- Position the AN/PAQ-4 on top of the bracket so that the spacer of the AN/PAQ-4 is aligned with the mounting knob of the bracket.
- Turn the mounting knob clockwise until the AN/PAQ-4 is tight.

WARNING
Do not store the AN/PAQ-4 with batteries installed.

(2) ***M136 (AT4) Launcher and Cartridge Dismounting Procedures***.

STEP 1: Detach the AN/PAQ-4 from the AN/PVS-4 mounting bracket.

- Turn the mounting knob counterclockwise until the AN/PAQ-4 is loose.
- Remove the AN/PAQ-4 from the bracket adapter.

STEP 2: Remove the adapter from the AN/PAQ-4

STEP 3: Remove the mounting bracket.

- Rotate the locking latch counter clockwise to disengage the latch shaft.
- Swing pivot bracket away from around the AT4.

F-2. AN/PEQ-2A TARGET POINTER/ILLUMINATOR/AIMING LIGHT

The AN/PEQ-2A target pointer/illuminator/aiming light (TPIAL) (Figure F-6) is a Class IIIb laser that emits a collimated beam of infrared light for precise aiming of the weapon as well as a separate infrared illuminating beam with adjustable focus. A safety block is provided for training purposes (blue side), which limits the operator from selecting the high power modes (black side). The TPIAL projects an infrared laser beam that cannot be seen with the eye but can be seen with night vision devices. It is also capable of projecting a much wider infrared illuminating beam from an integral illuminator. The TPIAL works with night vision goggles and mounts on various weapons with mounting brackets and adapters. The AN/PEQ-2A can also be used in the hand-held mode to illuminate and designate targets by leaders.

a. **Technical Data**.

Weight:	7.5 ounce (with 2 AA batteries)
Length:	16.26 centimeters (6.4 inches)
Width:	7.12 centimeters (2.8 inches)
Height:	3.05 centimeters (1.2 inches)
Range:	600 meters in low power (eye safe)
	2,000 meters in high power (non-eye safe)
Output Power:	
Aiming laser	25 m W (+- 10%)
Illuminator	30 m W (+50,-20%)
Beam Divergence:	
Aiming laser	.5 m rad
Illuminator	1.0-115 m rad
Batteries:	2 x 1.5V AA batteries

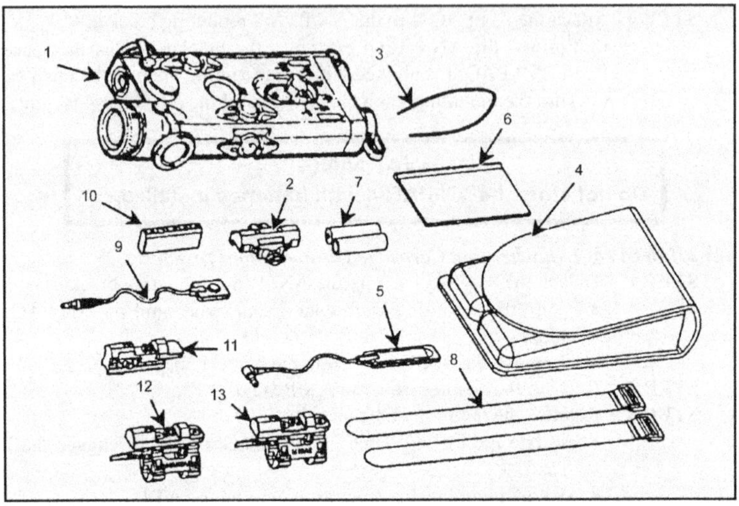

(1) TPIAL ASSEMBLY WITH SAFETY BLOCK
(2) RAIL
(3) NECK CORD
(4) TEXTILE BAG
(5) CABLE SWITCH, 12-INCH, MEMBRANE
(6) OPERATOR'S MANUAL
(7) BATTERIES 1.5V AA
(8) STRAP, RETENTION
(9) CABLE SWITCH, 20-INCH, BUTTON
(10) BRACKET ADAPTER
(11) TRAINING EXTENDER (ARMY ONLY)
(12) M4/M16A2 BRACKET ASSEMBLY

Figure F-6. AN/PEQ-2A with accessories.

b. **Operation.** This paragraph describes the battery installation, the safety block installation, the mode switch, the button switch, the cable switch, the focus switch, the lens cap, and the boresight adjusters.

(1) ***Battery Installation*** (Figure F-7).

(a) Unscrew the battery caps and install two AA batteries.

(b) Orient the batteries as indicated by the markings on the AN/PEQ-2A body.

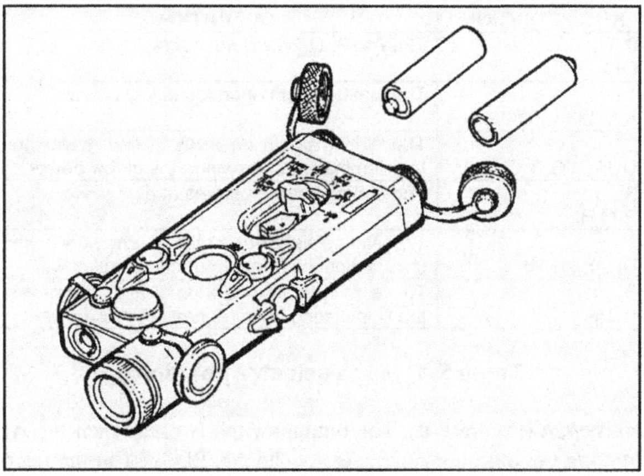

Figure F-7. AN/PEQ-2A battery installation.

(2) ***Safety Block Installation*** (Figure F-8). The safety block installed in the training mode (blue side up) prevents the operator from accessing the noneye safe modes (AIM HI, DUAL LO/HI, DUAL HI/HI). A .050 hex head Allen wrench is needed to unscrew the block from the body and reinstall it in the tactical mode (black side up).

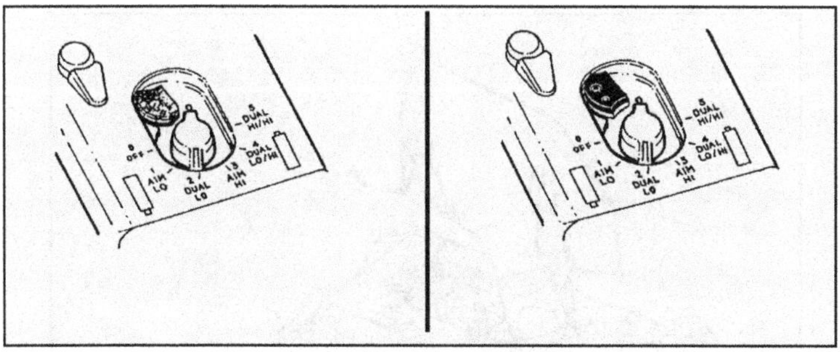

Figure F-8. Safety block installation.

(3) ***Mode Selector***. The mode selector is used to set the mode in which the AN/PEQ-2A operates when the cable switch button or push button are depressed. The mode selector has six positions (Table F-1).

KNOB POSITION	OPERATION
0 OFF	The AN/PEQ-2A will not operate.
1 AIM LO	The aiming beam operates at low power.
2 DUAL LO	The aiming beam operates at low power and the illuminating beam operates at low power.
3 AIM HI	The aiming beam operates at high power.
4 DUAL LO/HI	The aiming beam operates at low power and the illuminating beam operates at full power.
5 DUAL HI/HI	The aiming beam operates at high power and the illuminating beam operates at full power.

Table F-1. Mode selector positions.

(4) ***Button Switch*** (Figure F-9). The button switch is used when the AN/PEQ-2A is hand held. Pressing the button switch operates the AN/PEQ-2A in the operational mode set by the mode selector. When the button is released, the AN/PEQ-2A turns off.

(a) A green light emitting diode (LED) is incorporated into the body of the AN/PEQ-2A to indicate that the AN/PEQ-2A is ON. Whenever the AN/PEQ-2A is activated, the green LED will light and stay lit until the unit is turned OFF.

(b) If continuous operation of the AN/PEQ-2A is desired, pressing the button switch twice in rapid succession will latch the AN/PEQ-2A ON. The AN/PEQ-2A will remain on until the push button is pressed a third time.

Figure F-9. Operation of the button switch.

(5) ***Cable Switch*** (Figure F-10). The cable switch is used when the AN/PEQ-2A is mounted on a weapon. The cable switch plugs into the back of the AN/PEQ-2A

assembly. Pressing the button or pad at the end of the cable switch causes the AN/PEQ-2A to turn on in the operational mode selected by the mode select switch. When the button is released, the AN/PEQ-2A turns off.

(a) If continuous operation of the AN/PEQ-2A is desired, pressing the cable switch twice in rapid succession will latch the AN/PEQ-2A ON. The AN/PEQ-2A will remain on until the push button is pressed a third time.

(b) When the cable switch plug is installed in the AN/PEQ-2A, it automatically locks into place. To remove the switch, pull back on the plug sleeve and pull the plug out. DO NOT TRY TO REMOVE THE PLUG BY PULLING ON THE CABLE.

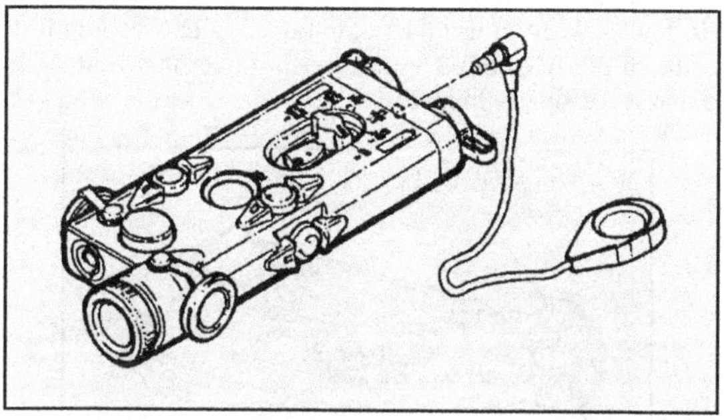

Figure F-10. Installation of the cable switch.

(6) ***Focus Knob*** (Figure F-11). The focus knob is used to vary the spread of the illumination beam based on the range and size of the area to be illuminated.

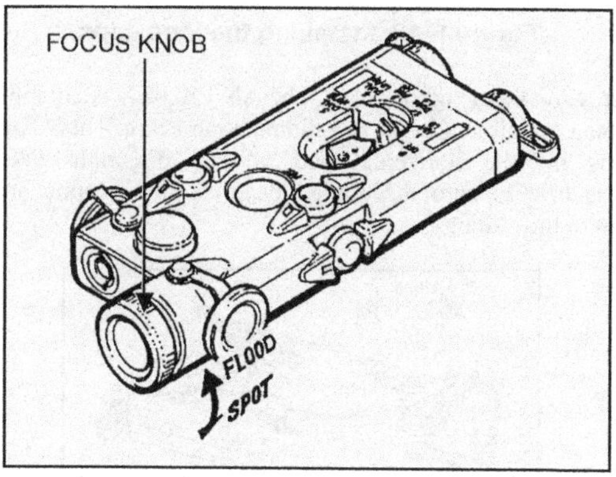

Figure F-11. Using the focus knob.

(7) ***Lens Cap*** (Figure F-12).

(a) The *black lens cap* blocks the AN/PEQ-2A illuminator or aiming laser beam should the device be activated. To use the black lens cap, pull it from its stored location on the side of the AN/PEQ-2A and stretch it over the front of the focus knob or aiming beam so that it fits snugly in place.

(b) The *diffuser lens cap* enables the illuminator or aiming laser to emit in a 45-degree cone (10 feet at 10 feet). To use the diffuser lens cap, pull it from its stored location on the side of the AN/PEQ-2A and stretch it over the front of the focus knob or aiming beam so that it fits snugly in place.

(c) The *neutral density lens cap* enables the AN/PEQ-2A illuminator or aiming laser to be operated in low power. To use the neutral density lens cap, pull it from its stored location on the side of the AN/PEQ-2A and stretch it over the front of the focus knob or aiming beam so that it fits snugly in place.

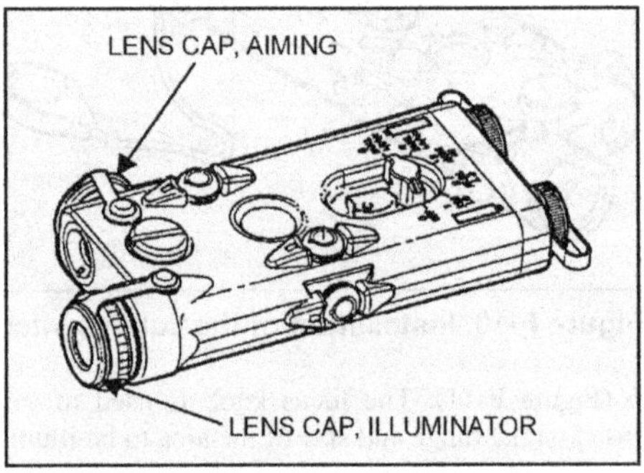

Figure F-12. Installing the lens caps.

(8) ***Boresight Adjusters*** (Figure F-13). The AN/PEQ-2A is equipped with boresight adjusters for zeroing the aiming beam and illumination beam. The AN/PEQ-2A adjusters move the beams in true horizontal and vertical directions. When zeroing the AN/PEQ-2A, it is best to zero the aiming beam to the weapon and then align the illumination beam to the aiming beam.

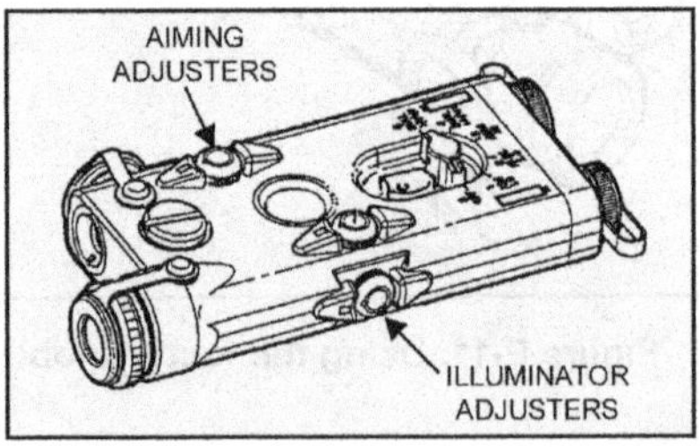

Figure F-13. Boresight adjusters.

F-3. AN/PAS-13 (V2) MEDIUM WEAPON THERMAL SIGHT and AN/PAS-13 (V3) HEAVY WEAPON THERMAL SIGHT

The AN/PAS-13 (V2) medium weapon thermal sight (MWTS) and the AN/PAS-13 (V3) heavy weapon thermal sight (HWTS) (Figure F-14) are silent, lightweight, compact, and durable battery-powered infrared imaging sensors that operate with low battery consumption.

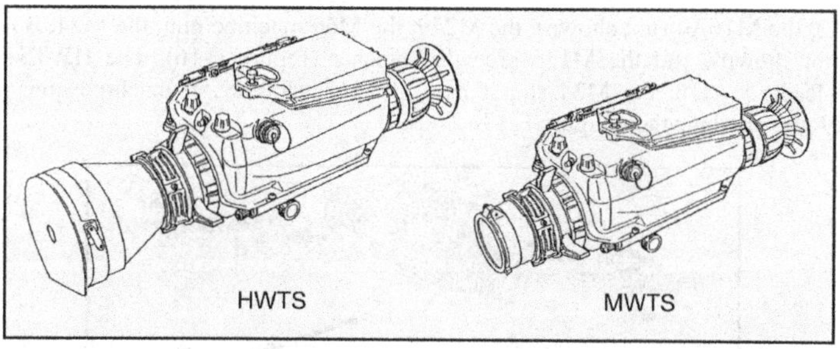

Figure F-14. HWTS and MWTS models of the thermal weapon sight.

a. **Description.**

(1) Both the MWTS and the HWTS are referred to as singular thermal weapon sights (TWS). The TWS is capable of target acquisition under conditions of limited visibility such as darkness, smoke, fog, dust, and haze. The TWS operates effectively at night and can also be used during the daytime. Infrared light is received through the telescope, detected by an IR sensor, converted to digital data, processed, and then displayed for the user. The TWS is composed of two functional groups: the telescope and the basic sensor.

(a) *Telescope.* The telescope receives IR light emitting from an intended target and its surroundings. The telescope magnifies and projects the IR light onto the scanner on the basic sensor (Figure F-15).

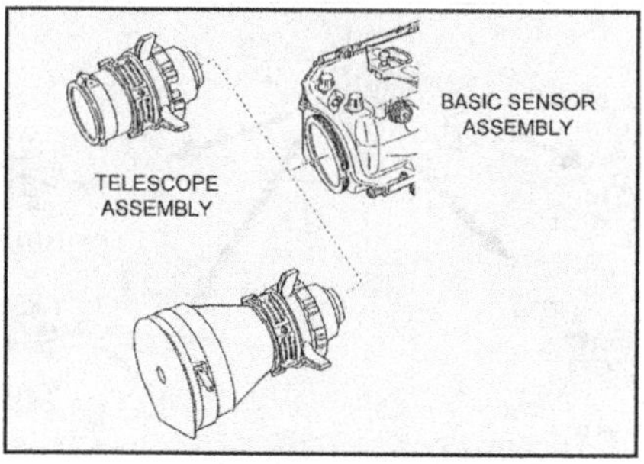

Figure 3-6. TWS configurations.

(b) *Basic sensor.* The scanner reflects the IR light received from the telescope onto the detective assembly. The detective assembly senses the IR light and converts it to video. The sensor's electronics condition the video for display on the LED array. The LED array illuminates the IR image along with the reticle. The light from the LED array is reflected off the scanner to form an image at the eyepiece.

(2) The only difference between the MWTS and the HWTS are the telescopes, which are different magnifications and hold different reticles. The basic sensor on the two models is the same. The MWTS fits the M4 carbine, the M4 modular weapon system, the M16A2, the M16A4 (not shown), the M249, the M60 machine gun, the M240B machine gun (not shown), and the M136 grenade launcher (Figure F-16). The HWTS fits the squad leader's M16, the M24 sniper rifle, the M2, caliber .50 machine gun, and the MK-19 grenade launcher (Figure F-17).

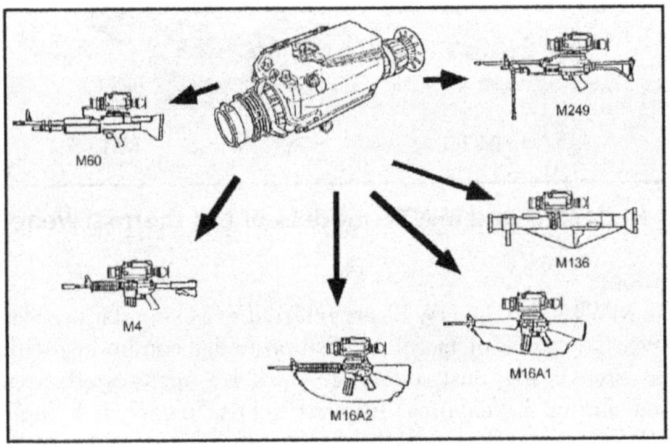

Figure F-16. MTWS.

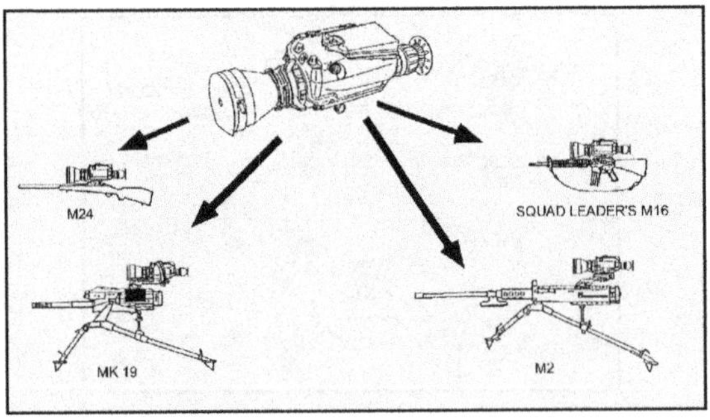

Figure F-17. HTWS.

b. **Technical Data.**

	MWTS	HWTS
Field of view (FOV):		
Narrow	9 degrees	3 degrees
Wide	15 degrees	9 degrees
Telescope magnification:		
Narrow FOV	3.3X	10X
Wide FOV	2X	3.3X
Length:	15.5 inches	18 inches
Width:	6.25 inches	6.25 inches
Height:	6.25 inches	6.25 inches
Weight	4.1 pounds	4.5 pounds
Power consumption:		
ON mode	12.0 watts	12.0 watts
STANDBY mode	5.5 watts	5.5 watts
Battery life:		
71% STANDBY mode	10 hours	10 hours
EMERGENCY mode	3.5 hours	3.5 hours

c. **Operation.**

(1) The TWS has three modes of operation: STANDBY, ON, and EMERGENCY.

(a) STANDBY mode. When the system is first turned on, the TWS begins a cool down period of approximately two minutes. After the cool down period, the TWS enters the STANDBY mode. During the STANDBY mode, power is not applied to the scanner or display in order to extend the life of the battery.

(b) ON mode. When the TWS is in the STANDBY mode and pressure is applied to the eyecup, the TWS switches to the ON mode, and a switch engages to provide power to the scanner and display. After a three-second delay, the system is fully operational.

(c) EMERGENCY mode. When switched to the EMERGENCY mode, the TWS continuously applies power to the entire system. This allows the operator to bypass the three-second delay experienced when switching from the STANDBY to the ON mode. Since power is applied to the entire system while in the emergency mode, battery life is greatly reduced.

(2) Controls and indicators are as shown in Figure 3-18.

(a) The BRIGHTNESS CONTROL is a nine-position rotary switch with an off detent position (turned fully counterclockwise). The purpose is to turn the system on or off and adjust the brightness of the eyepiece display.

(b) The CONTRAST CONTROL adjusts the contrast of the thermal image displayed on the raster. It has an automatic and a manual mode.

(c) The FOCUS RING adjusts the telescope focus from 20 meters to infinity. It requires a manual adjustment and affects both the wide and narrow fields of view.

(d) The FIELD OF VIEW RING is located on the telescope. It has a wide and a narrow field of view. The wide FOV is for using low magnification during target detection, and the narrow FOV is for using high magnification during recognition and engagement.

(e) The RETICLE SELECT SWITCH selects one of the available reticles depending on the TWS model (medium or heavy). It must be held for two seconds to enable reticle changes. After two seconds, release the switch to cycle to the next reticle. This control is disabled after ten seconds of inactivity.

(f) The RETICLE ADJUST SWITCH adjusts the reticle aiming features in azimuth and elevation. It is used during zeroing, and it must be held for two seconds to allow changes to be made. After two seconds, each press moves the reticle aiming features one increment. This control is also disabled after ten seconds of inactivity.

(g) The BLACK/WHITE POLARITY SWITCH selects the polarity of the thermal image displayed on the raster. The initial setting is "white hot." The polarity switch affects the appearance of the target.

(h) The DIOPTER FOCUS RING adjusts the focus of the raster and indicators to the operator's eye. It ranges from +2 to -6 diopters.

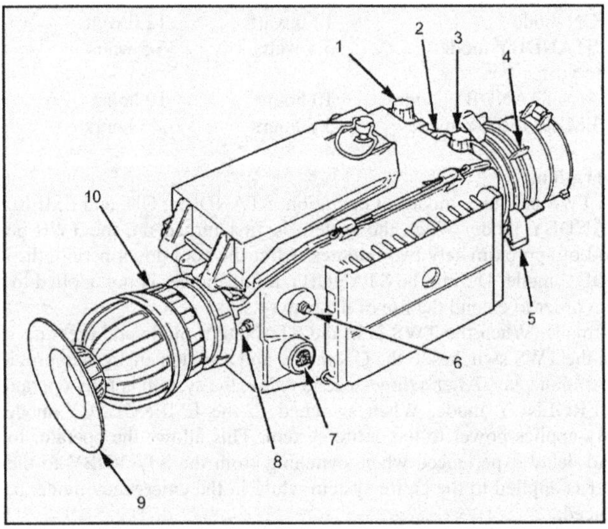

(1)	BRIGHTNESS CONTROL	(6)	RETICLE SELECT SWITCH
(2)	EMERGENCY SWITCH CONTROL	(7)	RETICLE ADJUST SWITCH
(3)	CONTRAST CONTROL	(8)	BLACK/WHITE POLARITY SWITCH
(4)	FOCUS RING	(9)	EYECUP
(5)	FIELD OF VIEW RING	(10)	DIOPTER FOCUS RING

Figure F-18. TWS controls and indicators.

(i) The EYEPIECE INDICATORS (Figure F-19) illuminate as follows:

- NOT COOL when the detectors are not cool enough for proper operation.
- WHT HOT/BLK HOT polarity.
- EMER during emergency mode operation.
- LOW when battery power has approximately fifteen minutes of useful power left.

(j) The COOLDOWN PERIOD INDICATORS blink the first ten seconds after the system is turned on. After ten seconds, the NOT COOL indicator is lit and the POLARITY, EMERGENCY, and LOW BATTERY indicators function normally.

(k) The DISPLAY RASTER displays the thermal image with a superimposed reticle.

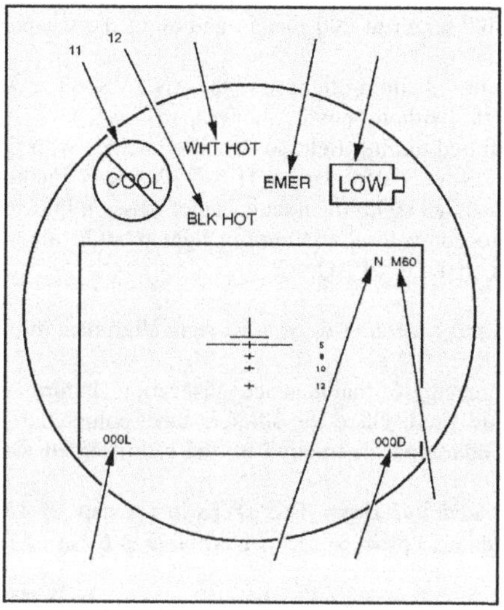

Figure F-19. Eyepiece indicators.

F-4. M136 AT4/M287 AT4 SUBCALIBER TRACER TRAINER SIGHT ALIGNMENT PROCEDURES

The fundamentals of marksmanship are the same as during the daytime with two exceptions. When using aided vision devices, the firer should hold the mounting bracket with the left hand. This method provides greater stability and helps to balance the AT4 against the added weight of the sight. When aiming the AT4, remember to aim by placing the infrared aiming light center mass on the target.

> **WARNING**
> **When firing the M136 AT4, do not place your face within 3 inches of the mounted sight. The recoil of the AT4 could cause injury you or damage to the equipment.**

a. Select a stable position for the weapon.

b. Open the M136 front and rear sight covers. During limited visibility conditions, the 7-millimeter peephole may be necessary for zeroing.

c. Set the rear sight for 200 meters and choose the appropriate scenario.

(1) Place a 25 meter zero target with a predetermined offset at 10 meters and align the weapon's fixed sight to coincide with the 4-centimeter aiming point of the zero target.

(2) Select a suitable target at 200 meters and align the weapon's fixed sight center mass of the target.

d. Turn on the infrared aiming light. Use the AN/PVS-7B or AN/PVS-14 to observe the infrared aiming light (without moving the weapon).

e. Adjust the infrared aiming light so that it coincides with the 2-centimeter offset aiming point of the 25-meter zero target. If the 200-meter method is used, adjust the aiming point of the mounted sight to coincide center mass on the 200-meter target.

f. Repeat the procedure until the aiming light is sight aligned to the weapon by adjusting the controls on the AN/PAQ-4C.

NOTE: Use the same procedures to conduct sight alignment of the AN/PEQ-2.

g. Conduct a preventive maintenance inspection before the instructional and qualification tables are fired. Once the soldiers have completed Tables I and III (day tables), they receive additional instruction on the use of aided vision devices for night operations.

(1) ***Table I (Instructional Day)***. This scenario consists of 12 targets from 100 to 300 meters. The standard to proceed to the next table is 6 out of 12 targets successfully engaged.

(2) ***Table II (Instructional Night)***. This scenario consists of 12 targets from 100 to 300 meters. The standard to proceed to the next table is 6 out of 12 targets successfully engaged at night.

(3) ***Table III (Qualification Day)***. This scenario consists of 10 targets from 100 to 300 meters. The standard to proceed to the next table is 8 out of 10 targets successfully engaged.

(4) ***Table IV (Qualification Night)***. This scenario consists of 10 targets from 100 to 300 meters. The standard to proceed to the next table is 6 out of 10 targets successfully engaged.

h. The night sustainment training strategy should be conducted over a one-day time frame and consists of the following:

(1) ***Night Vision Goggles Familiarization Training***. Soldiers receive instruction on the proper use and fit of night vision goggles, to include characteristics and capabilities, maintenance, and mounting procedures.

(2) ***Infrared Aiming Light Familiarization Training***. Soldiers boresight the weapon and aided vision device at a range of 10 meters.

(3) ***Fundamentals of Firing (Body Position with Head Harness and NVD, Foxhole and Prone)***. Soldiers review and practice firing positions and fundamentals of marksmanship and any changes that may occur by use of the infrared aiming device.

(4) ***Weapon and Equipment Precombat Inspection***. The weapon and infrared aiming device is inspected and properly mounted. Inspect to ensure that the infrared aiming

device is mounted securely to the mount and that the mount is tightly secured to the weapon. A review of clearing or misfire procedures is important.

(5) ***Infrared Aiming Light Boresight to Weapon***. Each soldier boresights the weapon and infrared aiming device at a range of 10 meters. If a bore light is not available, then zero the weapon and sight by using the specified zeroing procedure for that weapon/sight combination.

GLOSSARY

AC	Active Component
ANCOC	advanced noncommissioned officer course
AN/PVS	Army Navy passive vision sight
ARTEP	Army Training and Evaluation Program
backblast	The backward blast created by igniting the propellant in a weapon designed to be as recoilless as possible.
blast	An explosive or violent detonation that produces a direct effect on its target in addition to radiant heat and overpressure at its point of origin.
BMP	a type of non-US fighting vehicle built by the former Soviet Union
BNCOC	basic noncommissioned officer course
BRDM	a type of non-US scout car built by the former Soviet Union
BTR	a type of non-US personnel carrier built by the former Soviet Union
C	Centigrade
CS	combat support
CSS	combat service support
DA	Department of the Army
DKIE	decontamination kit individual equipment
DOD	Department of Defense
DS	direct support
F	Fahrenheit
FHT	field handling trainer

fps	feet per second
FTX	field training exercise
HE	high-explosive
HEAT	high-explosive antitank
HELLFIRE	heliborne laser fire and forget (a missile system)
HMTS	medium weapon thermal sight
HWTS	heavy weapon thermal sight
IAW	in accordance with
IET	initial entry training
IR	infrared
kmph	kilometers per hour
LAW	light antitank weapon
LED	light emitting diode
METL	mission-essential task list
MILES	Multiple Integrated Laser Engagement System
mm	millimeter(s)
MMTT	manned moving target tank
MOPP	mission-oriented protective posture
MOUT	military operations on urbanized terrain
mph	miles per hour
mps	meters per second
MTP	mission training plan

MWTS	medium weapon thermal sight
NA	not applicable
NATO	North Atlantic Treaty Organization
NBC	nuclear, biological, chemical
NCO	noncommissioned officer
NSN	national stock number
NVD	night vision device
OIC	officer in charge
overpressure	increased atmospheric pressure (positive overpressure), followed by a wave of decreased atmospheric pressure (negative overpressure), produced around the origin of an explosive or violent detonation
PLDC	primary leadership development course
POI	program of instruction
promethium	a radioactive metallic element of the rare-earth group used to illuminate range markings on the front sight of the M72-series LAW
RC	Reserve Component
RPG	rocket-propelled grenade (non-US)
STX	situational training exercise
SMCT	soldier's manual of common tasks
SOP	standing operating procedure
spall	small fragment or chip
STP	soldier training publication

STRAC	standards in training commission
TOC	tactical operations center
TP	training-practice
TPIAL	target pointer/illuminator/aiming light, AN/PEQ-2A
TRADOC	(US Army) Training and Doctrine Command
trc	track (as used in STRAC manual)
TRP	target reference point
TSC	training support center
TWS	thermal weapon sight

REFERENCES

SOURCES USED

These are the sources quoted or paraphrased in this manual.

TM 9-1055-886-12&P	Operator and Unit Maintenance Manual for M287 9-MM Tracer Bullet Training Device. 25 October 1989.
TM 9-1315-886-12	Operator's and Unit Maintenance Manual for Launcher and Cartridge, 84 Millimeter: M136 (AT4). 15 May 1990.
TM 9-1340-214-10	Operator's Manual for 66-MM Light Antitank Weapon (LAW) System M72A1, M72A2 with Coupler, M72A3, and Practice Rocket Launcher M190 with M73 Practice Rocket. 20 February 1992.
TM 11-5855-213-10	Operator's Manual for Night Vision Sight, Individual-Served Weapon AN/PVS-4. 1 February 1993.
TM 11-5855-213-23&P	Unit and Direct Support Maintenance Manual for Night Vision Sight, Individual-Served Weapon, AN/PVS-4. 1 June 1993.

DOCUMENTS NEEDED

These documents must be available to the intended users of this publication.

DA Form 7323-R	M72-Series LAW Scorecard. July 1994.
DA Form 7324-R	M136 AT4 Scorecard. July 1994.
DA Pamphlet 350-38	Standards in Weapons Training. 3 July 1997.
*DOD 5100.76-M	Physical Security of Sensitive Conventional Arms, Ammunition, and Explosives. 3 February 1983.
FM 7-7	The Mechanized Infantry Platoon and Squad (APC). 15 March 1985.
FM 7-7J	The Mechanized Infantry Platoon and Squad (Bradley). 7 May 1993.
FM 7-8	The Infantry Platoon and Squad. 1 March 2001.
FM 25-101	Battle Focused Training. 30 September 1990.

M72-SERIES LAW SCORECARD

For use of this form, see FM 23-25; the proponent agency is TRADOC

NAME ______________________ SSN ______________________ DATE __________

GRADE ______ RATING ______ UNIT ______________ WEATHER (circle one) CLEAR RAIN FOG SNOW

DATA REQUIRED BY THE PRIVACY ACT OF 1974

AUTHORITY: 10 USC 3012(g) / Executive Order 9397 **PRINCIPAL PURPOSE(S):** Record individual performance. **ROUTINE USES:** Evaluate individual proficiency and determine proficiency level. SSN is used for positive identification purposes only. **MANDATORY OR VOLUNTARY DISCLOSURE AND EFFECT ON INDIVIDUAL NOT PROVIDING INFORMATION:** Voluntary. Individuals not providing information cannot be rated/scored on a mass basis.

TABLE	TYPE OF FIRE	TYPE OF TARGET	ROUND	RANGE (M)	FIRING POSITION	HIT	MISS
1	Instructional DAY	Stationary	1	125 to 150	Standing	___	___
			2	150 to 175	Kneeling	___	___
			* 3	150 to 200	Sitting	___	___
			* 4	125 to 150	Prone	___	___
					Table Subtotals	___	___

FIRER'S SIGNATURE ______________________ DATE ________ SCORER'S SIGNATURE ______________________ DATE ________

TABLE	TYPE OF FIRE	TYPE OF TARGET	ROUND	RANGE (M)	FIRING POSITION	HIT	MISS
2	Instructional DAY	Moving	1	100 to 200	Standing	___	___
			2	100 to 200	Sitting	___	___
			* 3	100 to 200	Kneeling	___	___
					Table Subtotals	___	___

FIRER'S SIGNATURE ______________________ DATE ________ SCORER'S SIGNATURE ______________________ DATE ________

TABLE	TYPE OF FIRE	TYPE OF TARGET	ROUND	RANGE (M)	FIRING POSITION	HIT	MISS
3	Instructional NIGHT	Stationary	1	100 to 150	Kneeling	___	___
			2	100 to 150	Sitting	___	___
			* 3	100 to 150	Sitting	___	___
					Table Subtotals	___	___

FIRER'S SIGNATURE ______________________ DATE ________ SCORER'S SIGNATURE ______________________ DATE ________

* Firer must wear a protective mask when firing this round.

CONSOLIDATED SCORES

FIRER'S SIGNATURE ______________________ DATE ____________ TOTAL MISSES ________

SCORER'S SIGNATURE ______________________ DATE ____________ RANK ________

RANGE OIC'S SIGNATURE ______________________ DATE ____________ RANK ________

INSTRUCTOR NOTES	
ALL FIRING TABLES	• Do not announce the ranges to targets. • Fire all engagements in sequence.

DA Form 7323-R, JUL 94

M136 AT4 SCORECARD

For use of this form, see FM 23-25; the proponent agency is TRADOC

NAME ______________________ SSN ______________________ DATE __________

GRADE ______ RATING ______ UNIT ______________ WEATHER (circle one) CLEAR RAIN FOG SNOW

DATA REQUIRED BY THE PRIVACY ACT OF 1974

AUTHORITY: 10 USC 3012(g) / Executive Order 9397. **PRINCIPAL PURPOSE(S):** Record individual performance. **ROUTINE USES:** Evaluate individual proficiency and determine proficiency level. SSN is used for positive identification purposes only. **MANDATORY OR VOLUNTARY DISCLOSURE AND EFFECT ON INDIVIDUAL NOT PROVIDING INFORMATION:** Voluntary. Individuals not providing information cannot be rated/scored on a mass basis.

TABLE	TYPE OF FIRE	TYPE OF TARGET	ROUND	RANGE (M)	FIRING POSITION	HIT	MISS
1	Instructional Day	Stationary	1	150 to 200	Kneeling	___	___
		Stationary	* 2	150 to 200	Kneeling	___	___
		Stationary	3	250 to 300	Sitting	___	___
		Stationary	4	250 to 300	Sitting	___	___
		Stationary	5	125 to 150	Prone	___	___
		Stationary	6	125 to 150	Prone	___	___
		MMTT/MAS	7	150 to 200	Kneeling	___	___
		MMTT/MAS	* 8	150 to 200	Kneeling	___	___
		MMTT/MAS	9	200 to 250	Standing	___	___
		MMTT/MAS	*10	200 to 250	Fighting position, unsupported	___	___
		MMTT/MAS	11	250 to 300	Kneeling	___	___
		MMTT/MAS	12	250 to 300	Kneeling	___	___
					Table Subtotals	___	___

FIRER'S SIGNATURE ______________________ DATE __________ SCORER'S SIGNATURE ______________________ DATE __________

TABLE	TYPE OF FIRE	TYPE OF TARGET	ROUND	RANGE (M)	FIRING POSITION	HIT	MISS
2	Instructional Night	Stationary	1	125 to 150	Prone	___	___
		Stationary	2	125 to 150	Prone	___	___
		Stationary	* 3	150 to 200	Sitting	___	___
		Stationary	4	150 to 200	Sitting	___	___
		Stationary	5	250 to 300	Kneeling	___	___
		Stationary	6	250 to 300	Kneeling	___	___
		MMTT/MAS	7	125 to 150	Standing	___	___
		MMTT/MAS	8	125 to 150	Standing	___	___
		MMTT/MAS	* 9	200 to 250	Kneeling	___	___
		MMTT/MAS	10	200 to 250	Kneeling	___	___
		MMTT/MAS	11	150 to 200	Standing	___	___
		MMTT/MAS	12	150 to 200	Fighting position, unsupported	___	___
					Table Subtotals	___	___

FIRER'S SIGNATURE ______________________ DATE __________ SCORER'S SIGNATURE ______________________ DATE __________

* Firer must wear a protective mask when firing this round.

INSTRUCTOR NOTES	
ALL FIRING TABLES	• Do not announce the ranges to targets. • Fire all engagements in sequence. • Fire Table 1 before Table 3; fire Table 2 before Table 4.
FIRING TABLES 2 THROUGH 5	• Ensure no one fires at front or rear of a manned moving tank target. • Total hits and misses, and enter results in the consolidated scores block on the back of this scorecard.

DA Form 7324-R, JUL 94 (FRONT)

INDEX

FIELD MANUAL
NO. 3-22.34(23-34)

HEADQUARTERS
DEPARTMENT OF THE ARMY
Washington, DC, 28 November 2003

TOW WEAPON SYSTEM

CONTENTS

DISTRIBUTION RESTRICTION: Approved for public release; distribution is unlimited.

*This publication supersedes FM 23-34, 17 August 1994.

PREFACE

This manual discusses the many changes in the TOW missile, TOW training, and OPFOR armored vehicles and countermeasures. It includes training information on the M220A1 (basic) and M220A2 tube-launched, optically tracked, wire-guided (TOW) weapon systems and carrier M966. Its intended audience includes leaders and members of mechanized infantry, light infantry, airborne, and airmobile infantry organizations who train and test individuals and crews. Units with M2/M3 Bradley fighting vehicle systems should also refer to FM 23-1 for additional training and testing requirements.

The proponent of this publication is US Army Infantry School. Send comments and recommendations on DA Form 2028 directly to Commandant, US Army Infantry School, ATTN: ATSH-INB-B, Fort Benning, Georgia 31905-5595, or email to doctrine@benning.army.mil.

Unless otherwise stated, masculine nouns and pronouns do not refer exclusively to men.

CHAPTER 1
INTRODUCTION

The TOW is a crew-portable heavy antitank missile weapon system that can be operated from armored, lightly armored, and unarmored multipurpose vehicles and helicopters, and from the ground mount. The system consists of a guided missile and a launcher to engage tanks, fortifications, and other materiel targets. The term "TOW" is an acronym derived from the words that describe the principal operating features of the missile—tube-launched, optically tracked, wire-guided. The launcher initiates, tracks, and controls the flight of the missile by guidance signals transmitted over a command-link wire that connects the missile with the launcher.

1-1. CHARACTERISTICS

The TOW weapon system consists of a launcher, which has tracking and control capabilities, and the missile, which is encased in a launch container. The launcher is equipped with self-contained, replaceable units.

a. **Capabilities.** The TOW is mainly an antitank weapon used for long-range engagement of armored targets. It can be employed in all weather conditions as long as the gunner can see his target through the daysight tracker or nightsight. The TOW also provides a long-range assault capability against heavily fortified bunkers, pill boxes, and gun emplacements.

(1) The ITOW missile can destroy targets at a minimum range of 65 meters and a maximum range of 3,750 meters. The TOW 2B missile can destroy targets at a minimum range of 288 meters when fired from the ground mount and 200 meters when fired from the HMMWV or BFV. The TOW 2B has a maximum range of 3,750 meters whether ground- or vehicle-mounted.

(2) The automatic missile tracking and control capabilities of the TOW weapon system provide a high first-round-hit probability. To operate the system, the gunner places the crosshairs of the sighting system (either the daysight tracker or the nightsight) on the target, fires the missile, and centers the crosshairs on the target image until missile impact. The optical tracking and command functions within the system guide the missile along the gunner's line of sight. The gunner does not apply lead, windage, or elevation.

b. **Mobility.** The TOW weapon system can be vehicle-mounted or ground emplaced (tripod-mounted) for operation. Missiles can be launched from either operational mode. (Detailed operating procedures for each configuration are provided in Chapters 2 and 3.)

(1) The entire system can be hand-carried by a weapon crew; therefore, emplacement sites can be changed quickly to minimize detection or to engage targets that are out of range.

(2) The vehicle-mounted launcher is more mobile and can be quickly prepared for use. The launcher can be assembled and disassembled without the use of tools.

(3) Built-in self-test circuits can be used to check the operational condition of the assembled launcher whether ground- or vehicle-mounted.

1-2. SYSTEM CONFIGURATIONS

The TOW weapon system comes in two configurations—the M220A1 (TOW) and the M220A2 (TOW 2).

a. **M220A1 TOW Weapon System.** The M220A1 TOW (Figure 1-1) consists of a tripod, a traversing unit, a launch tube, a daysight tracker, an AN/TAS-4 nightsight, a missile guidance set (MGS), a battery assembly housed in the compartment of the MGS, and an encased missile.

(1) The M220A1 TOW weighs about 265.5 pounds with all of its components and carrying cases; with encased missile (BGM-71A), it weighs 320 pounds. (For a detailed description of the M220A1 TOW, see TM 9-1425-472-12.)

(2) The M220A1 TOW can fire all four configurations of TOW missiles, which include many types. However, it cannot take full advantage of the BGM-71D TOW 2 and BGM-71E TOW 2A or TOW 2B counter countermeasures when tracking through obscurants.

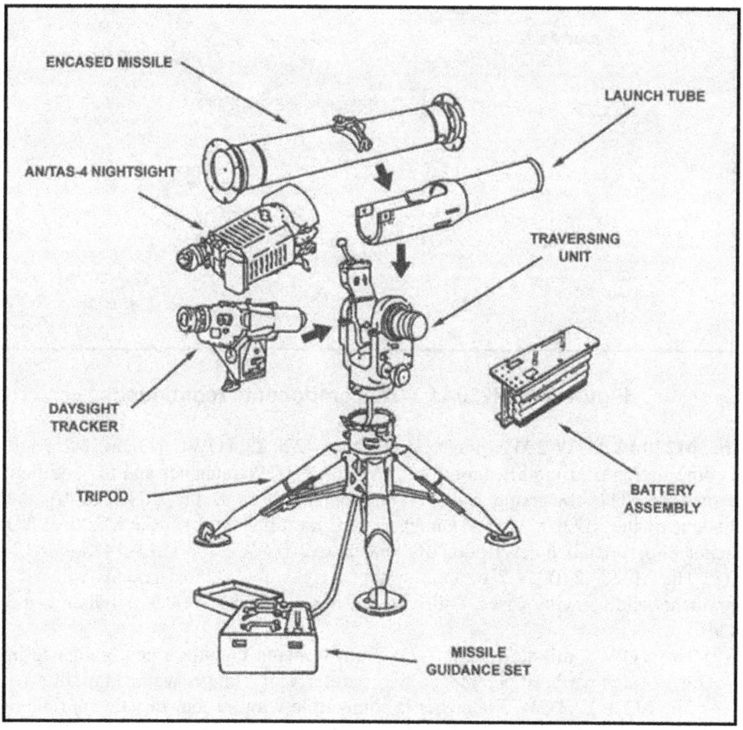

Figure 1-1. M220A1 TOW components.

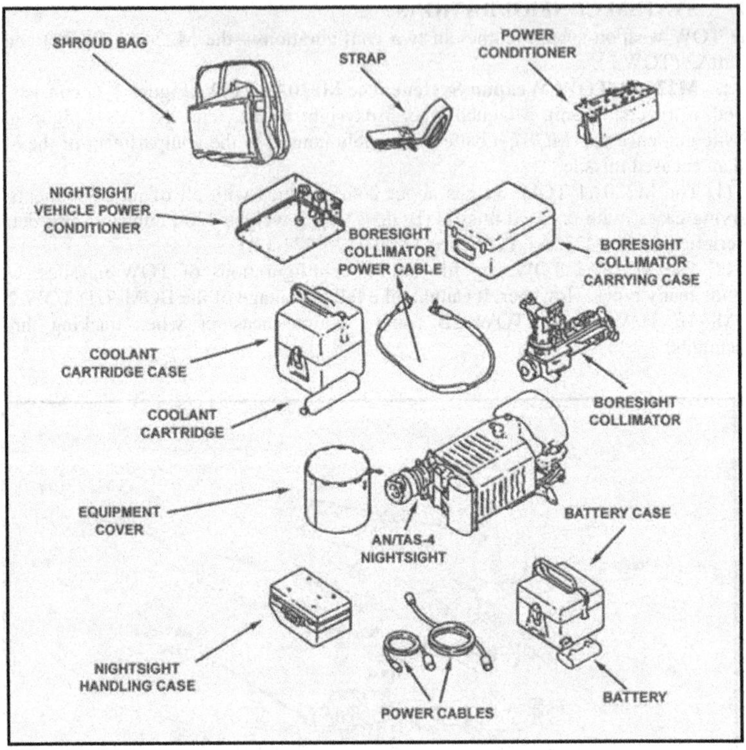

Figure 1-1. M220A1 TOW components (continued).

b. **M220A2 TOW 2 Weapon System.** The M220A2 TOW 2 (Figure 1-2, page 1-4) is a combination of a modification to the M220A1 TOW launcher and the addition of a new missile. The traversing unit, the digital MGS, and the AN/TAS-4A thermal nightsight of the M220A1 TOW launcher were modified to form the M220A2 TOW 2 launcher. (For a detailed description of the M220A2 TOW 2, see TM 9-1425-450-12.)

(1) The M220A2 TOW 2 weapon system weighs about 256.5 pounds with all of its components and carrying cases; with encased missile BGM-71D, it weighs about 318.5 pounds.

(2) The TOW 2 missile (BGM-71D) improvements include a new guidance link, a full-caliber 6-inch warhead, a reloaded flight motor, and a longer warhead probe.

(3) The M220A2 TOW 2 launcher is compatible with all four missile configurations. It can achieve a higher probability of hit against all types of targets through improved microprocessor-based electronics that use digital design techniques.

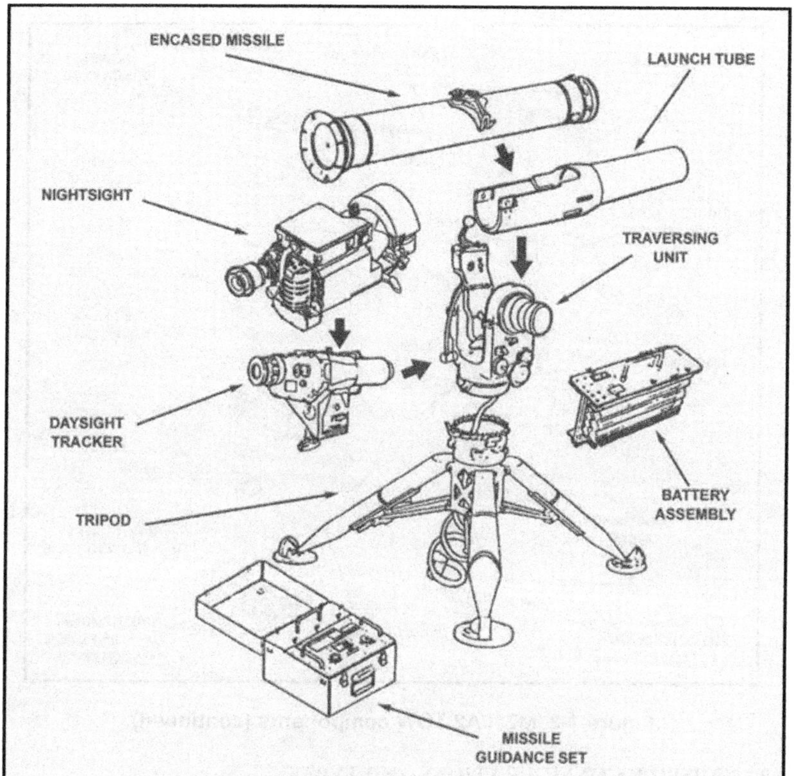

Figure 1-2. M220A2 TOW components.

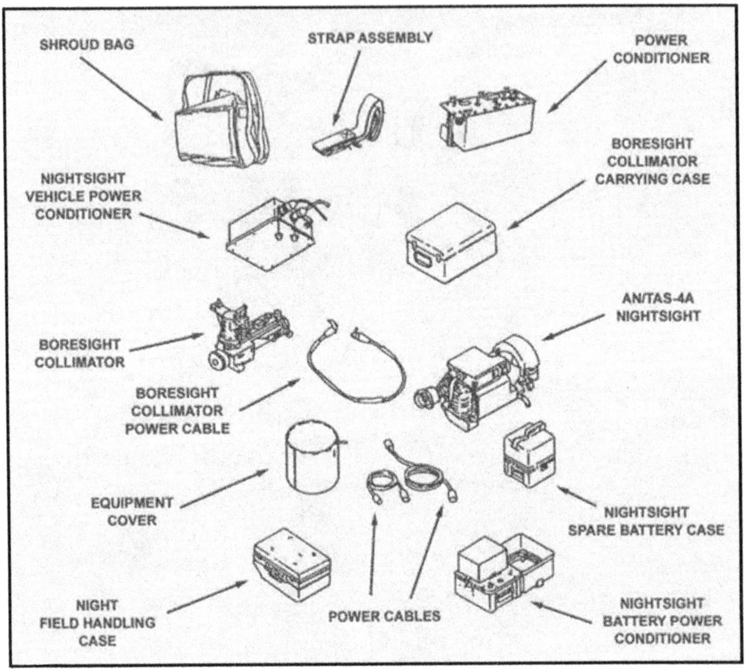

Figure 1-2. M220A2 TOW components (continued).

1-3. MISSILE CONFIGURATIONS AND TYPES

The TOW missile comes in four configurations with numerous types. These types mainly consist of minor modified work orders (MWOs) that are transparent to the operator and are continually added to. This FM will not attempt to deal with the various types but will confine itself to the four major configurations. All configurations use the same basic airframe, aerodynamic control system, command-link wire, and missile electronics designs.

a. The first configuration is the improved TOW (ITOW). These missiles have improved 5-inch warheads that include extended probes for greater standoff and penetration.

b. The second configuration is the TOW 2. This missile has a full-caliber 6-inch warhead that includes an extended probe. In addition to the infrared radiator of the ITOW missile, TOW 2 has a second infrared radiator to provide hardened system performance against battlefield obscurants and countermeasures. The second radiator is called the thermal beacon and provides link compatibility with the electrooptical infrared nightsight, which is part of the TOW 2 launcher system.

c. The third configuration is the TOW 2A. The TOW 2A has an added small explosive charge in the tip of the extended probe providing improved performance against reactive armor by premature detonation.

d. The fourth configuration is the TOW 2B. The TOW 2B has an entirely different warhead and kill mechanism than the other TOW missiles. It is a top-attack missile (fly over/shoot down [FOSD]) that defeats enemy armor at its most vulnerable point—the top deck of the turret and hull. The TOW 2B has a tandem warhead that fires two explosively formed projectiles (EFPs) down through the thin upper deck armor of the enemy vehicle. The gunner tracks the target the same as any other TOW missile with the crosshairs on center mass, but the missile automatically flies 2.25 meters above the line of sight (LOS). When the missile senses that it is directly above the target (by means of the target's shape and magnetic field), it automatically fires its warhead.

1-4. TOW MISSILE IDENTIFICATION

With the increased number of TOW missile types and wide variations in their capabilities (especially armor penetration), TOW crews and range safeties must be able to quickly and accurately identify the specific type of missile they have.

a. The configuration of the missile can be readily identified by the various identification decals, tapes, and stencils on the missile case (Figure 1-3 and Table 1-1). (See TM 9-1410-470-34 for current information on TOW missiles.)

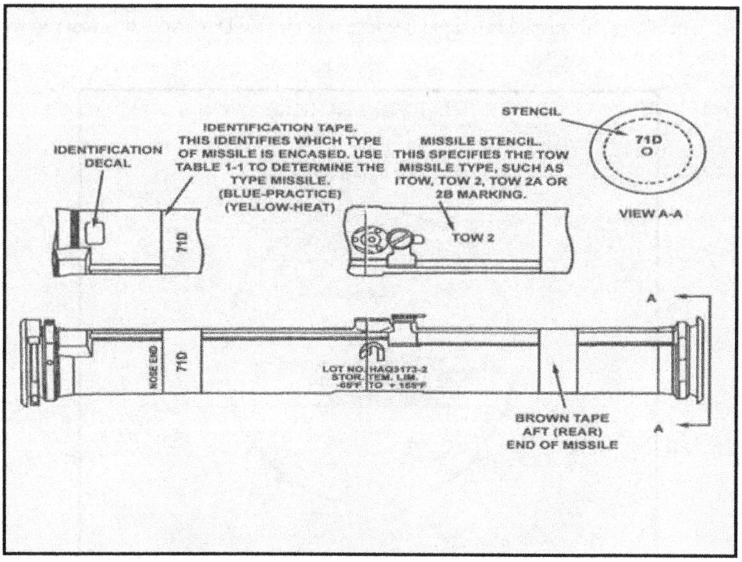

Figure 1-3. TOW missile case markings.

TYPE OF MISSILE	DECAL PLATE DATA
Improved TOW Practice	Guided Missile Practice, STD, BTM-71A1
Improved TOW HE	Guided Missile Surface Attack, ITOW-BGM-71C
TOW 2 Practice	Guided Missile Practice, TOW 2, BTM-71D-1B
TOW 2 HE	Guided Missile Surface Attack, TOW 2, BGM-71D-3B
TOW 2A Practice	Guided Missile Practice, TOW 2A, BGM-71E-3B
TOW 2B Practice	Guided Missile Practice, TOW 2B, BTM-71E-2B
TOW 2B HE	Guided Missile Surface Attack, TOW 2B, BGM-71F
Decals are located on the nose end of the launcher tube. They describe the type of round and model. BLUE—indicates practice round. YELLOW—indicates HE round. WHITE—indicates HE or practice round. BROWN—indicates low explosive (live rocket motor).	

Table 1-1. TOW missile identification.

b. The TOW 2B missile is unique because it is an FOSD missile. It is marked with a decal as shown in Figure 1-4.

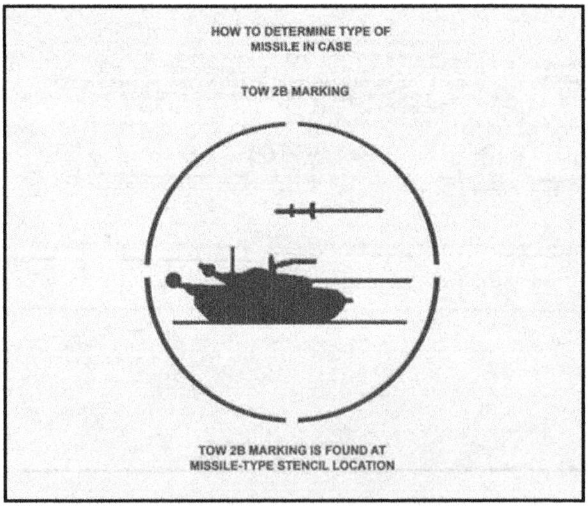

Figure 1-4. TOW 2B marking.

1-5. INTERNAL COMPONENTS AND SEQUENCE OF OPERATIONS

A TOW crewman or leader who thoroughly understands his weapon will be able to employ it to its maximum effectiveness. This paragraph describes the internal missile operations from trigger depression to missile impact, which will increase the soldier's knowledge of how his weapon functions.

a. **Major Internal Sections and Components.** The missile is divided into three major sections and the launch container.

(1) ***Front Section.*** The front section contains the warhead and the electronics.

(a) The warhead contains an extended probe on the ITOW and TOW 2-series (17 inches on TOW 2, 10 inches on ITOW). All missile types have a crush ogive switch to detonate the missile (located in the end of the probe tip on ITOW and TOW 2). The warhead is a conventional shaped charge (except TOW 2B) with a copper cone and liner and a detonator at the rear of the charge. The diameter of the warhead is 5.2 inches on ITOW and 6 inches on TOW 2 and 2A.

(b) All electronic circuitry necessary to perform on-board electronic functions while the missile is in flight is contained in the front section. This circuitry receives steering signals from the MGS and corrects the missile flight path to conform to LOS.

(2) ***Center Section.*** The center section contains the thermal batteries, the gyroscope, the flight motor, and wings.

(a) The missile has three thermal batteries, which are used instead of storage batteries because their shelf life is much longer. The chemical reaction that produces both heat and electricity is started by an electrical charge that is part of the prefire signal sent when the trigger is depressed. Within approximately two tenths of a second the batteries begin producing electricity.

(b) The gyroscope stabilizes the missile in flight, and it provides information about the attitude of the missile to the MGS from sensors in its housing. When the prefire signal is sent to the missile, an explosive squib on the neck of a bottle of compressed nitrogen is ignited. This produces the popping sound and whirring noise that is heard after the gunner depresses the trigger and before the missile launches. The escaping nitrogen causes the gyroscope to begin spinning in its bracket. When the gyroscope reaches 42,000 rpm, it automatically disengages from its bracket and begins to spin freely in its housing. This causes an electrical current that ignites the launch motor launching the missile from the weapon system.

NOTE: The process of the thermal batteries heating up to produce electricity and the gyroscope coming up to speed and igniting the launch motor takes about 1.5 seconds. This causes the delay between the time when the gunner presses the trigger and the missile actually fires.

(c) The flight motor is a solid fuel rocket motor with two nozzles protruding through the body of the missile at 30-degree angles. The flight motor burns for 1.6 seconds. During this time, it provides all the momentum necessary for the full flight of the missile.

(d) For the rest of its flight, the missile coasts on momentum and the lift provided by the four wings. The wings are located just aft of the flight motor, recessed into the body of the missile during storage and spring loaded. When the missile leaves the launch container, the wings open and lock into place. The wings are set at 45-degree angles to the body of the missile.

(3) ***Aft Section.*** The aft section contains the wire spools, the launch motor, the actuator control system and control surface push rods, a bottle of compressed helium, the flight control surfaces, the xenon beacon (on all missiles), and the thermal beacon (on TOW 2-series missiles only).

(a) *Wire Spools.* Two wire spools, each holding 3,750 meters of wire, are located at the rear of the missile. The wire is made of fine steel and is coated with a thin layer of varnish for insulation. There is virtually no tension on the wire, so it droops down on the ground behind the missile as the missile flies downrange.

(b) *Launch Motor.* The launch motor is a solid fuel rocket that burns entirely inside the launch tube. It provides just enough thrust to propel the missile to a point where the flight motor can ignite safely without injuring the gunner.

(c) *Flight Control Surfaces.* Four flight control surfaces are located on the aft end of the TOW missile set at 90-degree angles to the body of the missile. The TOW, unlike the Dragon, does not spin in flight but maintains a stable position. One pair of flight control surfaces controls the "up and down" (pitch) corrections while the other pair controls the "left and right" (yaw) corrections. Like the wings, the flight control surfaces are recessed into the body of the missile while in the launch container. They spring out and lock into place when the missile leaves the launch container.

(d) *Actuator Control System/Push Rods/Helium Bottle.* The TOW missile makes steering corrections by moving the flight control surfaces. It moves them by opening and closing the valve on a bottle of compressed helium. The escaping helium moves a system of push rods back and forth. The push rods are connected to the flight control surfaces and move them in turn. The valve on the helium bottle opens and closes in response to signals sent by the on-board electronics, which are responding to signals sent down the wires by the MGS.

(e) *Xenon Beacon.* The xenon beacon emits an infrared beacon, which is picked up by the infrared receiver on the daysight tracker. The infrared receiver measures the angle at which the infrared beam strikes, thus providing the major source of data on the position of the missile to the MGS. The xenon beacon consists of a bulb filled with xenon gas and two electrodes. When an electric current passes between the two electrodes, it creates a spark, which excites the gas. This excited gas emits infrared light that exits from a window on the rear of the beacon as a narrow beam. All basic TOW infrared beacons operate on the same frequency (the infrared light turns on and off at the same rate of speed). This causes two problems: First, two systems cannot be placed closer than 300 meters because the beacons overlap and the MGS has no means of distinguishing between the missiles, causing it to lose control. Second, jamming the daysight tracker is fairly simple if the enemy knows the correct frequency. The TOW 2 overcomes these problems by having the MGS send a signal to the missile that controls the frequency at which the xenon beacon is operating. The MGS varies this pattern randomly, speeding it up and slowing it down in no apparent pattern. The MGS is always able to distinguish its missile from other missiles because no two missiles will be operating on the same frequency at the same time. For the same reason, the enemy cannot jam the system.

(f) *Thermal Beacon.* Under some battlefield conditions such as heavy smoke, dust, or fog, the xenon beacon, which operates on the lower end of the infrared spectrum, cannot penetrate the obscuration. With the TOW missile and weapon system it is sometimes possible for the gunner to be able to acquire the target, yet the infrared receiver on the daysight tracker cannot acquire the infrared beacon and the system loses the missile. TOW 2-series missiles have a thermal beacon to correct this deficiency. The thermal beacon operates on the upper end of the infrared spectrum and will penetrate any obscurant the nightsight can see through. A postamplifier was added to the AN/TAS-4A nightsight on the TOW 2 weapon system to acquire the thermal beacon and track it in much the same manner as the infrared beacon. The basic AN/TAS-4 is a passive sight, but the AN/TAS-4A is an active sight with the ability to steer the missile to any target that the gunner can see through the sight itself.

(4) ***Launch Container.*** The launch container is the fiberglass tube that holds the missile. It protects the missile from environmental and physical damage while in storage, secures it in the weapon system when about to be fired, and electrically connects the missile to the weapon system during firing.

b. **Sequence of Operations.** The sequence of operations (Table 1-2) starts from the time that the gunner presses the trigger. The time that each event occurs is given in negative or positive numbers, with 0 being the time that the launch motor fires.

TIME (SECONDS)	EVENT
-1.5	Gunner presses trigger; system sends prefire sequence to missile, starts chemical reaction in thermal batteries, blows explosive squib off nitrogen bottle; gyroscope starts spinning.
0	Gyroscope reaches 42,000+ rpm and disengages; sends electrical current to launch motor; launch motor fires.
+0.5	Missile exits launch tube.
+0.8	Wing 2 switch closes.
+.10	Wing 4 switch closes; helium bottle opens; timed part of arming sequence starts; flight motor ignites.
+.145	Delay switch changes state (part of arming sequence).
+.18	Safety and arming unlock occurs. As missile accelerates to 19 Gs, the "G" force causes a rotor in the warhead to unlock.
+.53	S&A clock mechanism rotates and aligns detonator with warhead. Missile is armed.
+1.6	Flight motor burns out. Missile is fully armed and travelling at maximum velocity. It will coast the rest of the way on momentum and the lift from the four wings.

Table 1-2. Sequence of operations.

c. **Tracking and Steering of the Missile While in Flight**. As the missile flies downrange, the infrared beam from the rear of the missile enters the infrared receiver on the daysight tracker, is reflected by a spinning mirror in the rear of the infrared receiver called a nutater, and strikes the infrared sensor in the front of the infrared receiver. The pattern that this reflected beam makes on the sensor measures the angle of the beam and tells the MGS where the missile is. The MGS has two other sources of information on the location and attitude of the missile: the gyroscope and the traversing unit. Both of these have sensors. The gyroscope measures the attitude of the missile while in flight, and the traversing unit measures how quickly and in what direction the gunner moves it. The MGS uses this information to make a very precise determination of where the missile is in relation to the gunner's aiming point. This automatically steers the missile toward that point. This process continues until the missile impacts the target.

1-6. TOW TRAINING STRATEGY

TOW training strategy synchronizes institutional and unit training to produce TOW crews and units that win on the battlefield. (See Appendix A for training tips and Appendix B for a train-the-trainer program.)

a. The training strategy for TOW begins in 11B OSUT where the soldier is taught critical Skill Level 1 common skill tasks and critical 11B Skill Level 1 tasks. He is trained to be a loader, gunner, and driver.

b. Much individual and most collective, tactical, and gunnery training takes place in the unit. Primary sources for tactical and leader training are FM 7-91, ARTEP 7-91-MTP, and ARTEP 7-91-Drill.

c. The Skill Level 1 and 2 soldiers, squads, and crews are trained in the unit. Gunners take part in an individual gunner's qualification program that includes TOW Tables 1 and 3 and then qualify on Tables 2 and 4. The squads and crews practice on TOW Tables 5 and 7 and then qualify on Tables 6 and 8. (See Chapter 4 for more detailed information on TOW training. See DA Pam 350-381 for strategies and resource requirements.)

d. The section is the basic TOW fighting organization; therefore, section training is critical. Sections practice on TOW Table 9 and qualify on Table 10.

e. Platoon gunnery events consist mainly of Table 11 for practice and Table 12 for qualification. Also, the platoon must conduct a semiannual force-on-force exercise to demonstrate its proficiency with a maneuver element

f. TOW training in units is supported by MOS-specific tracks in BNCOC and ANCOC, and the Antiarmor Leader Course for NCOs (SGT[P] through SFC).

g. Army Correspondence Course Program (ACCP) courses also support TOW training. (See ACCP catalog for specific courses).

CHAPTER 2

TRIPOD-MOUNTED TOW/TOW 2

Regardless of the type of carrier the TOW weapon system is mounted on, at times it will be employed as a ground-mounted system. This may be due to tactical requirements, such as helicopter insertion for an antiarmor ambush, or it may be due to mechanical failure or battle damage to the carrier. The following information and procedures apply to all TOW units. During combat or field training, TOW crews will sometimes be required to dismount the TOW system. Each TOW crew dismounts the TOW IAW dismount procedures for the appropriate vehicle as outlined in this manual.

Section I. OPERATION PROCEDURES

Operation procedures for the TOW weapon system include performing a system checkout, conducting preventive maintenance checks and services (PMCS), selecting firing positions, and preparing TOW missiles for operation.

2-1. SYSTEM CHECKOUT

The TOW weapon system is set up on the tripod to conduct a system checkout to ensure that the system is working properly before engaging targets. System checkout procedures are conducted before and after installation of an M220-series launcher on its organic carrier; daily, after the system is in use; or anytime the system has been in storage for one week. (See TM 9-1425-472-12 [M220A1] or TM 9-1425-450-12 [M220A2] for complete details.) After the system checkout procedures are completed, the weapon is mounted or installed on its organic carrier IAW the appropriate TM.

2-2. PREVENTIVE MAINTENANCE CHECKS AND SERVICES

A preventive maintenance checks and services (PMCS) should be performed before, during, and after operation of the TOW weapon system. PMCS are performed daily if the system is in continuous use. (See TM 9-1425-472-12 [M220A1] or TM 9-1425-450-12 [M220A2] for complete details.)

2-3. SELECTION OF FIRING POSITIONS

When the TOW is assembled for ground emplacement, the tripod must be level for effective operation of the weapon. If the tripod is not level, lateral (azimuth) movement of the traversing unit produces a change in the elevation angle of the launch tube; therefore, acquiring and tracking the target may be difficult. An unleveled tripod can cause an electronic control error during the final tracking operation (while the command-link wire is guiding the missile to its target). For ground emplacement, proper adjustment of the tripod legs compensates for uneven ground up to a slope of 30 degrees. Therefore, the firing position must not slope more than 30 degrees. When the tripod legs are properly positioned, the bubble in each level will be between the marks on the level indicator. The TOW is not fired from a position where the traversing unit has a lateral cant of more than 10 degrees. The ideal firing position is as level as possible. However, the missile can be fired from a position where the tripod is canted as much as 30 degrees directly downhill or uphill. (See Appendix C for more information on fighting positions.) After the site is selected and the weapon emplaced, the squad leader and gunner make a range card for the position. (For detailed information on making a range card, see Appendix D.)

2-4. PREPARATION OF MISSILES

The TOW missile requires complete attention to detail when preparing it for firing missions. Missile identification, inspection, and correct handling procedures eliminate safety hazards to personnel and equipment. (See TM 9-1425-450-12 or TM 9-1425-472-12 for complete details.)

Section II. PERFORMANCE PROBLEMS AND DESTRUCTION PROCEDURES

TOW crew members must recognize and overcome performance problems and limitations to survive in battle. Each unit develops and maintains an SOP for the proper destruction of ammunition and weapon systems.

2-5. MALFUNCTIONS

A malfunction is the failure of ammunition to function correctly when fired, or the unexpected functioning of explosive components during a test. Premature functioning during handling, maintenance, storage, transportation, or deployment constitutes a malfunction. Any abnormal functioning such as failure of the missile to leave the launch container on launch motor ignition also constitutes a malfunction. During operation of the TOW, a malfunction occurs if the missile fails to launch after the normal 1.5-second delay after the firing trigger has been pressed or if the missile fails to function properly in flight. Two major types of malfunctions are *hangfire* and *misfire*. In-flight malfunctions are classified as two basic types—duds and erratic flight. Defective missiles (either fired or encased) must be disposed of by an explosive ordnance disposal (EOD) team only.

NOTE: When appropriate, weapon and vehicle TM checklists should be used when correcting malfunctions. (See TM 9-1425-450-12.)

a. **Hangfire**. A hangfire is an unexpected delay in the functioning of a series of ammunition components. It may occur anywhere in the series of events, from pressing the firing trigger to launch motor ignition. After the trigger is pressed, the distinctive impulse of the squibs firing (to activate the missile batteries and gyro) and the gyro spin noise can be heard.

(1) If the missile fails to fire, alert the crew by announcing "Hangfire" and continue to track the target for one minute.

(2) If the missile fails to launch after one minute, close the trigger cover, lower the arming lever, place the elevation lock in the 8-degree down position, and place the azimuth lock in the locked position (do not lock the azimuth if it points the missile out of the left and right limits). Open the bridge clamp and remove the missile from the launch tube. Carry it to a dud pit or a safe distance from friendly troops.

(3) Point the missile downrange at all times and notify EOD personnel.

WARNING

The encased missile will become hot to the touch if the TOW thermal batteries, which generate temperatures higher than 800 degrees F, are activated. The encased missile will produce smoke if the thermal beacon has been ignited.

b. **Misfire.** A misfire (complete failure to fire) occurs if the launch motor fails to ignite after the firing trigger has been pressed. This type of failure may be caused by a faulty firing circuit, a failure of electrical power, poor electrical connections, short circuits, or faulty components in the ignition or propellant systems. When the missile fails to launch, the prefire functions are not heard and the condition is a misfire. The target is tracked for one minute and the trigger is pressed again.

(1) Alert the crew by announcing, "Misfire," and check the battery (see notes 2 and 3). Then, check the MGS to ensure the coil cable connector is properly attached to connector J1. Turn the coil cable connector locknut clockwise until it is tight to make sure the coil cable connector is locked in place. Ensure the bridge clamp locking handle is properly seated. Lower the arming lever and check the backblast area to ensure no personnel are in danger; then, raise the arming lever. Press the trigger again.

(2) If the missile still fails to launch, continue tracking the target for one minute or at the gunner's discretion (in combat). Close the trigger cover, lower the arming lever, place the elevation lock in the 8-degree down position, and place the azimuth lock in the locked position. Open the bridge clamp and remove the missile from the launch tube.

(3) Move the missile a safe distance away (minimum 100 meters) and place it on the ground, preferably in a dud pit. Point the missile downrange at all times and notify EOD personnel.

c. **Dud**. A dud is any missile that has been launched (fired from the launch tube) but failed to fly to the target or failed to detonate on target impact. The warhead of all duds must be considered to be armed, and therefore, extremely dangerous. The location of all duds will be noted and referred to EOD personnel for disposal.

d. **Erratic Flight.** An erratic or uncontrolled missile flight may result due to any of the following conditions:

- Shorted or broken command-link wires.
- Defective missile wing or control surface.
- Defective rocket motor.
- Excessively high gunner tracking errors.

NOTES:
1. In a training situation, wait 30 minutes after announcing, "Hangfire" before proceeding.
2. For the M220A1, (MGS self-test position 1) check for proper band reading. If out of band reading, replace the battery.

3. For the M220A2, do MGS self-test for battery. If battery fails, replace the battery.
4. In a training situation, lower the arming lever, close trigger cover, place elevation lock in locked position, place azimuth lock in locked position, and wait 30 minutes.
5. Report all misfires, hangfires, and other missile malfunctions IAW AR 75-1.

2-6. FIRING LIMITATIONS

Some conditions may limit the firing and engagement capabilities of the TOW. The following information should be considered before engaging targets. (See TM 9-1425-450-12 for updated firing limitations.)

a. **Firing Over Water.** Firing across bodies of water wider than 1,100 meters can reduce the range of the TOW. Signals being sent through the command-link wires are shorted out when a large amount of wire is submerged in water. Maximum and limited range firing over water varies according to missile type. If the range is less than 1,100 meters, the missile's range is not affected. A TOW position should be as high above and as far back from the water as the tactical situation allows. The squad or section leader should analyze his sector as soon as the position is occupied to determine if water will affect the employment of the TOW.

(1) The various types of TOWs listed may be fired over water to a maximum range as shown in Figure 2-1 (TOW launcher and target at water surface level).

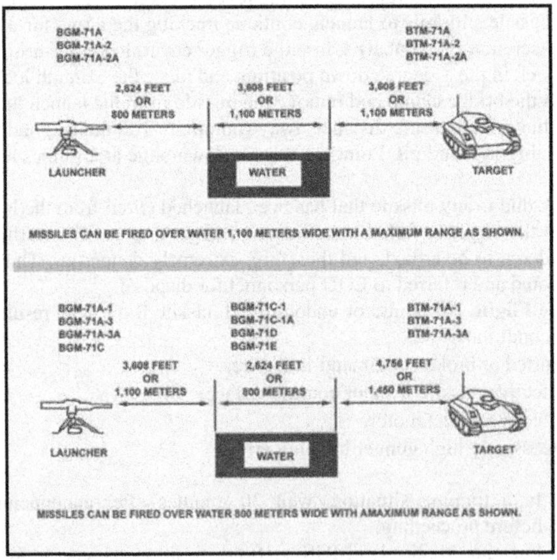

Figure 2-1. Maximum TOW missile range.

(2) The various types of TOWs listed may be fired over water to a limited range as shown in Figure 2-2 (TOW launcher and target at water surface level).

(3) To determine the extended range of missiles when firing over water (when TOW launcher or target is above surface level of water), use Table 2-1 (pages 2-6 and 2-7).

(a) Determine the type of missile being used.

(b) Determine the height above the water surface of the TOW launcher and place a tick mark at the corresponding height in the left-hand column of the table.

(c) Determine the height above the water surface of the target. Place a tick mark at the corresponding height in the right-hand column of the table.

(d) Connect the two tick marks with a straight line. The point where the lines intersect the center column is the maximum range the missile can travel without getting too close to the water.

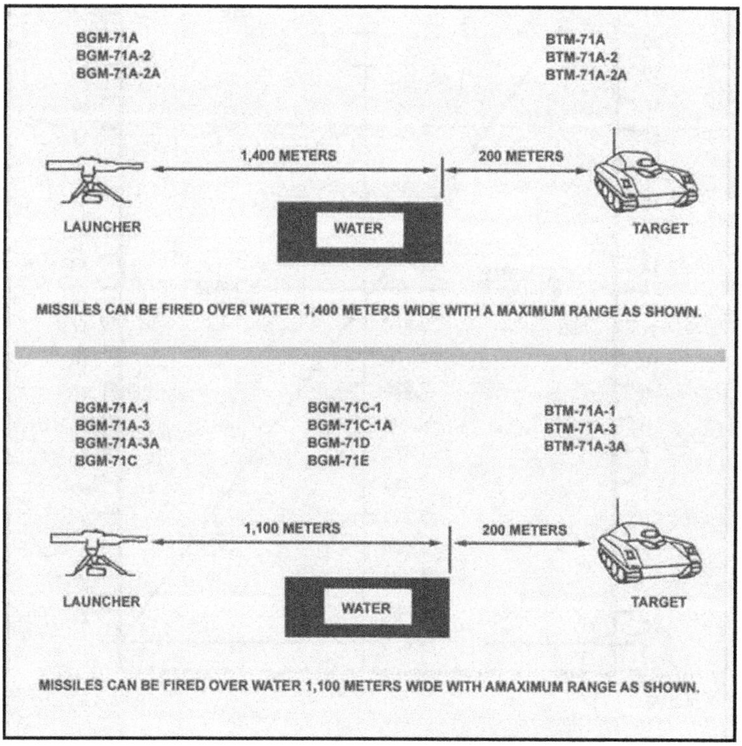

Figure 2-2. Limited TOW missile range.

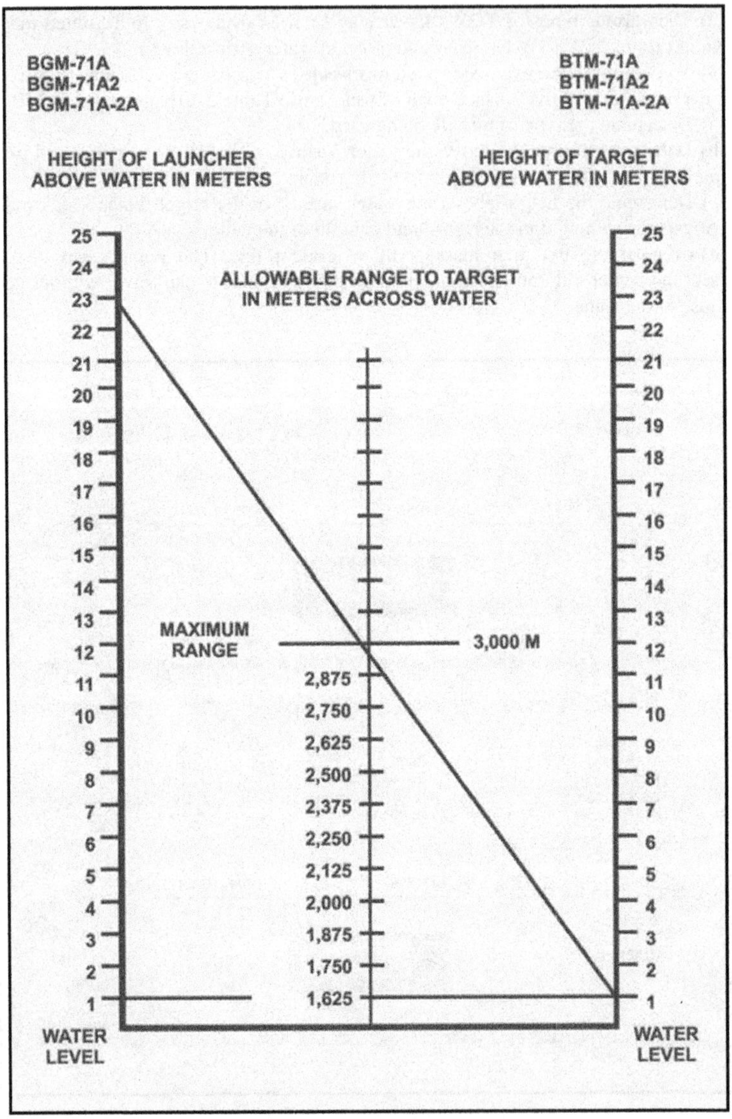

Table 2-1. Extended range over water.

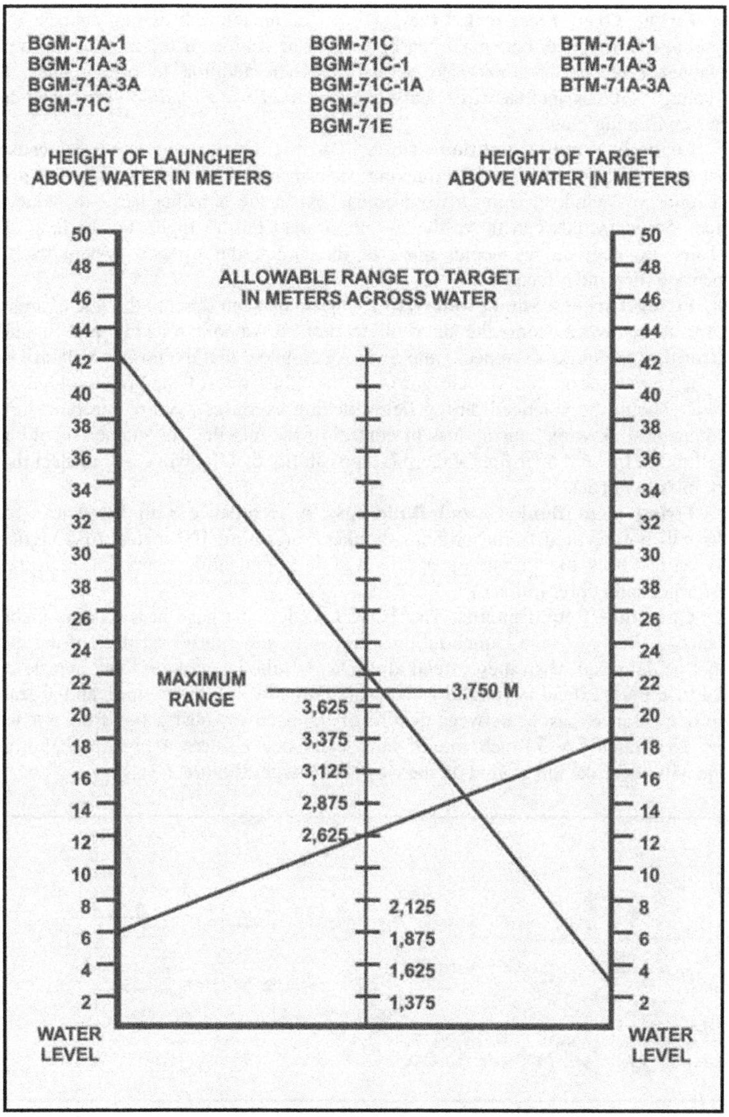

Table 2-1. Extended range over water (continued).

b. **Firing Over Electrical Lines.** If the command-link wires contact a live high-voltage power line, personnel can be injured or control of the missile can be lost. The launcher electronics may also be damaged. In addition to power lines, other high-voltage sources include street cars, electric train ways, and some moving target trolleys on training ranges.

c. **Firing in Windy Conditions.** Gusty, flanking, or quartering winds can cause the launch tube to vibrate and spoil the tracking performance. The effect is similar to driving in a strong crosswind. Erecting a windscreen next to the launcher helps to reduce this problem. Strong winds can move the missile around during flight, but as long as the crosshairs are kept on the center mass of the target, the weapon system itself can compensate for wind effects.

d. **Firing Through Smoke and Area Fires.** Smoke can obscure the line of sight and hide the target when using the daysight tracker. A smooth tracking rate should be maintained as the target disappears into a smoke cloud so that the missile will still be on target or very close as the vehicle goes out the other side of the smoke cloud. (This technique should be practiced during field tracking exercises.) A fire can burn through the command-link wire, causing loss of control of the missile. The gunner should avoid firing through fire and over fires if there is a possibility that the wires will contact the fire before missile impact.

e. **Firing from Bunkers and Buildings.** In accordance with DA Pam 385-63, TOWs will not be fired from buildings, bunkers, or within 100 meters of a vertical or nearly vertical backstop without the approval of the commanding general. (See Appendix E for further safety precautions.)

f. **Clearance Requirements.** The TOW muzzle must have at least nine inches of clearance at the end of the launch tube so the wings and control surfaces of the missile will not be damaged when they extend after clearing the launch tube. The muzzle of the launch tube must extend beyond any enclosure, window sill, or aperture, and at least 30 inches of clearance must be between the line of sight and any obstruction from 500 to 900 meters downrange. A 30-inch line-of-sight clearance ensures a high probability the missile will not strike the ground on the way to the target (Figure 2-3).

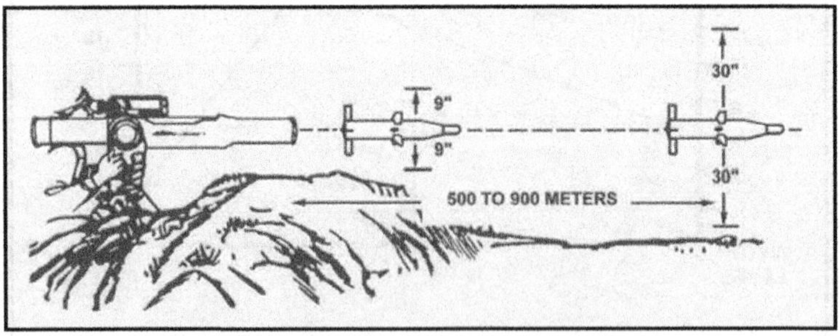

Figure 2-3. Clearance requirements.

2-7. DESTRUCTION OF TOW WEAPON SYSTEMS

When subject to capture in the combat zone, the encased missile and weapon system and its components are destroyed by the TOW squad. This action is taken when, in the judgment of the unit commander, the tactical situation warrants it, and when it is necessary IAW orders or policy established by the commander.

NOTE: The following information is for guidance only. Certain procedures outlined require the use of explosives and incendiary grenades that may not be authorized items for the TOW squad. The issue of these and related materials and the conditions for destruction are command decisions.

Destruction of TOW components by mechanical means, explosives, gunfire, or burning renders them useless to the enemy. The best method to destroy everything is to set fire to the carrier. To prevent the enemy from obtaining useful equipment, the encased missiles and certain weapon system parts should be destroyed. When lack of time or personnel prevents destruction of all parts, the encased missiles and the priority parts of each weapon system must be destroyed. Destroying the same parts of each weapon system prevents the enemy from assembling a complete weapon system. Priority is used to select the parts for destruction. If destruction is directed, the TOW squad selects a point of destruction that will cause the greatest obstruction to enemy movement and not create a hazard to friendly troops.

Each organization and installation that uses, maintains, or stores the encased missiles or weapon system parts should have an SOP for TOW destruction. The procedure should contain priorities of destruction, methods of destruction, quantities of explosives required, and instructions for destruction. The destruction plan should be flexible to cover any situation.

a. **Destruction of Encased Missiles.** The encased missile is a priority item and should be destroyed by any of the following methods. The expended launch containers should also be destroyed or removed from the battlefield so that enemy troops may not use them to make improvised munitions and booby traps.

WARNING

Do not attempt to destroy encased missiles by mechanical methods.

(1) ***Launch the Missiles.*** If the system is operational and time permits, launch the missiles into enemy territory before destroying the vehicle. Missile detonation should hinder the enemy in every way possible without endangering friendly units or equipment.

(2) ***Use Explosives.*** To destroy the TOW using explosives, become familiar with the provisions of FM 5-250.

(a) Using three pounds of TNT or equivalent, prepare and place a 1-pound explosive charge 6 to 8 inches from the nose end, a 1-pound explosive charge 4 to 6 inches behind the electrical connector, and a 1-pound explosive charge 6 to 8 inches from the motor end of the missile (Figure 2-4, page 2-10).

NOTE: Because the missile contains high explosive, a smaller explosive charge may be used to destroy it.

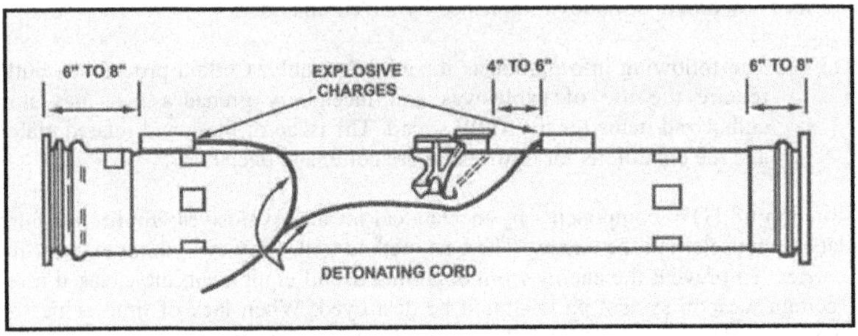

Figure 2-4. Explosive charges on an encased missile.

(b) Determine if electrical blasting caps and wire or nonelectric blasting caps and safety fuses will be used for priming and detonating the explosive charges. If nonelectric caps are used, crimp them to a safety fuse at least 2 meters long.

WARNING

Safety fuse burns at the rate of 1 foot in 30 to 40 seconds. Safety fuse, which contains black powder, and blasting caps must be protected from moisture at all times.

(c) Connect the charges with detonating cord to produce a simultaneous detonation. Dual prime the charges to reduce the possibility of a misfire.

WARNING

The blasting caps, detonating cord, and safety fuses must be kept separate from the charges until needed.

(d) If the charges are primed with nonelectric blasting caps, ignite the safety fuses and take cover at once. If the charges are primed with electric blasting caps, take cover before firing.

(3) ***Use Gunfire.*** Use well-aimed shots from artillery, rockets, rifle grenades, or tank guns to destroy the encased missile by gunfire.

WARNING

The live rocket motors and high-explosive antitank warhead are a hazard to soldiers firing at the encased missile. Therefore, small arms should not be used. Firing artillery from ranges of 1,000 meters or less must be done from cover. Cover must also be used when rifle grenades or rockets are fired.

b. **Destruction of Weapon Systems.** Destruction of TOW parts should be done so that the use or restoration to serviceable condition is impossible. In addition, certain weapon system parts could provide important information to the enemy.

(1) ***Priority of Destruction.*** If time does not permit the destruction of the complete weapon system, destroy as many parts as possible in the following priority:

- Daysight tracker.
- Nightsight.
- Missile guidance set.
- Traversing unit.
- Battery assembly.
- Launch tube.
- Tripod.
- Battery power conditioner (TOW 2).

(2) ***Destruction Methods.*** Consider the location of friendly units and the availability of personnel, materials, and time required for destruction when selecting the method of destroying TOW parts. The following list of destruction methods is in priority sequence; however, the considerations stated above could change the priority or method of destruction used.

(a) *Explosives.* Use at least one pound of explosive on each part. Place the explosive inside the part, if possible, to cause the greatest destruction. For example, remove the coil cable and place the charge in the traversing unit, or remove the battery assembly from the MGS, place the charge in the battery well, and replace the battery. After placing the charge, cover the parts with filled sandbags or other heavy materials to increase the explosive effect.

(b) *Gunfire.* Use well-aimed shots from artillery, rockets, tank guns, rifle grenades, rifle fire, or other small-arms fire to make the weapon system useless to the enemy. When using small-arms fire, aim for critical parts. The vehicle can be destroyed by artillery or tank guns from a range of 1,000 meters.

(c) *Mechanical.* Smash the TOW parts with axes, picks, crowbars, or rocks, or drive over each part with a tracked vehicle (make sure the parts are on a hard surface or they must be dug up to ensure they have been made useless). Destroy the priority parts if enough time and personnel are available.

WARNING

Do not try to destroy the battery assembly by mechanical methods. An electrical safety hazard may result if the power output terminals are short-circuited. Discard the battery assembly if no other means of destruction is available.

(d) *Burning.* After maximum damage has been done to the weapon system and vehicle parts by other destruction methods, and if time permits, set fire to the resulting debris. Use vehicle fuels and lubricants to aid burning. Place one or more incendiary grenades on each part.

CHAPTER 3
M966-MOUNTED TOW/TOW 2

The M966-mounted TOW is a one-vehicle (1 1/4-ton truck) combat system that is air-transportable, versatile, maintainable, and survivable (Figure 3-1). Its 16-inch ground clearance, four-wheel independent suspension, steep approach and departure angles of 60-percent gradation, side slope of 40 percent, and 30-inch (without kit, 60-inch with kit) water-fording capability provides the off-road mobility and speed needed in combat. The vehicle carries one complete launcher system, six encased missiles, and a three-man crew. This chapter discusses the duties of the crew and the detailed operations of the weapon system.

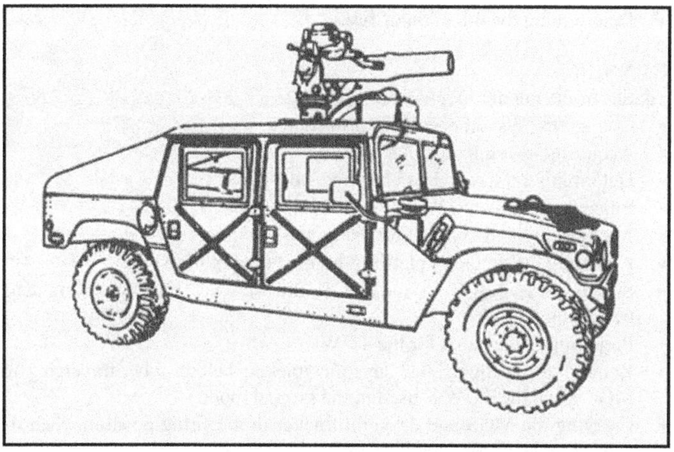

Figure 3-1. M966-mounted TOW.

Section I. INDIVIDUAL DUTIES AND RESPONSIBILITIES

The three-man crew consists of the squad leader, the gunner, and the driver-loader. The duties and responsibilities of each crew member are discussed in the following paragraphs.

3-1. SQUAD LEADER

The squad leader is responsible for the combat readiness and tactical employment of his squad. This includes discipline, health and welfare, training, and maintenance of equipment and weapons. Other specific duties include the following:

- Directing the driver over the exact route.
- Detecting and identifying targets.
- Issuing fire commands.

- Controlling fire and movement.
- Supervising and assisting in dismounting and assembly of the M220-series TOW launcher.
- Employing the squad according to the orders of the section leader.
- Selecting primary, alternate, and supplementary firing positions.
- Requesting squad resupply.
- Informing the section leader of events that affect the tactical situation and of the status of his crew.
- Carrying the tripod, launch tube, coolant cartridges, battery case or BPCs, and binoculars when the TOW is used in the ground mode.
- Acting as the gunner when needed.
- Assisting the gunner in system check-out procedures.
- Supervising construction of the TOW firing position.
- Determining the direction of fire.

3-2. GUNNER

Specific duties of the gunner include the following:

- Conducting system check-out procedures.
- Acquiring, recognizing, and identifying armored vehicles.
- Determining if a target can be engaged.
- Engaging targets.
- Recognizing and eliminating firing-angle limitations.
- Adjusting the gunner's platform before stowing the MGS.
- Stowing the TVPC in the MGS and stowing the MGS, nightsight, and traversing unit.
- Preparing range cards for the TOW.
- Removing the nightsight, daysight tracker, launch tube, traversing unit, and MGS when the TOW is used in the ground mode.
- Carrying the MGS and daysight tracker to the firing position when the TOW is used in the ground mode.

3-3. DRIVER-LOADER

The driver-loader is a dual-purpose position in the M966 TOW squad. Specific duties of the driver-loader include the following:

- Driving the M966.
- Maintaining the vehicle.
- Camouflaging the vehicle.
- Carrying the traversing unit, encased missile, nightsight, and collimator to the firing position when the TOW is used in the ground mode.
- Securing the area.
- Stowing battery assembly, collimator, encased missiles, daysight tracker, spare battery case, coolant cartridges and batteries (or BPCs), and tripod on the M966 with the assistance of the squad leader.
- Maintaining radio equipment.
- Removing misfired missiles.

Section II. OPERATION PROCEDURES

Before using the M966-mounted TOW/TOW 2, the squad must install the M220-series TOW launcher and encased missiles and must prepare the M966 for TOW firing. (To install an M220A1 TOW launcher and encased missiles on an M966, see TM 9-1425-472-12. To install an M220A2 TOW launcher and encased missiles on an M966, see TM 9-1425-450-12.)

3-4. EQUIPMENT STORAGE

The TOW system components are stored in the interior of the vehicle (Figure 3-2 [M220A1], page 3-4 and Figure 3-3 [M220A2], page 3-5).

a. Six encased missiles are stowed in the missile racks on the left rear of the cargo area.

b. The traversing unit is stowed on the traversing unit adapter behind the right front passenger seat.

c. The MGS is stowed on the gunner's platform between the left and right rear passenger seat.

d. The nightsight and collimator are stowed on the right cargo shelf.

e. The launch tube is stowed on the right side of the cargo shell door.

f. The BPC and spare battery pack, or the nightsight coolant cartridges and the 4.8-volt nickel cadmium battery, are stowed in the floor bracket next to the right wall of the cargo shelf.

g. The daysight tracker is stowed in the floor bracket next to the missile rack.

h. The battery assemblies are stowed in the battery rack above the missile rack.

i. The tripod is stowed retracted, lying across the rear of the cargo area.

j. The TVPC is stowed in the MGS.

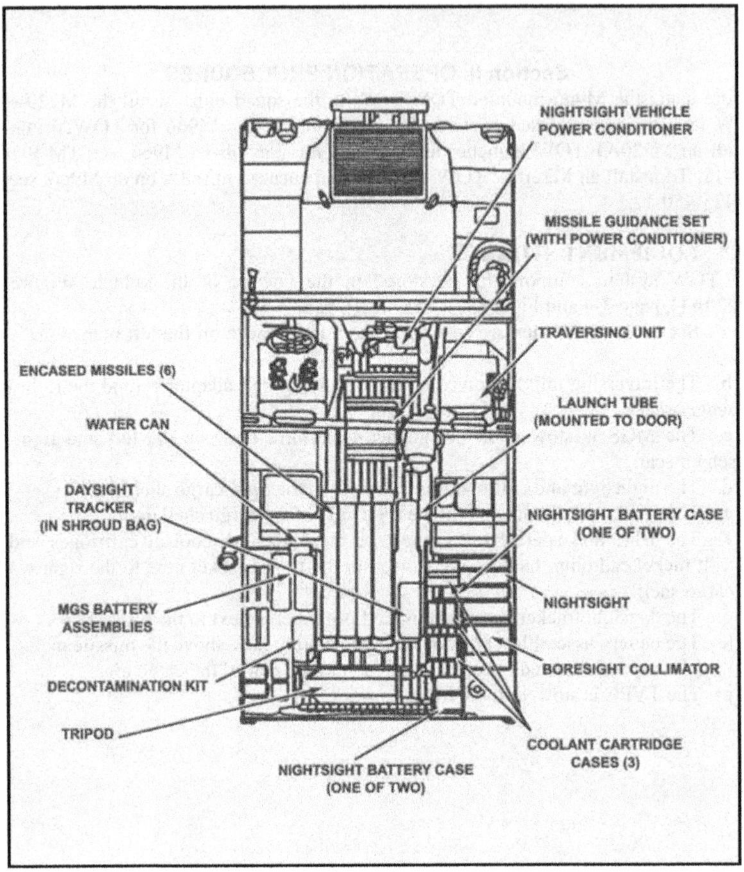

Figure 3-2. Stowing M220A1 on M966.

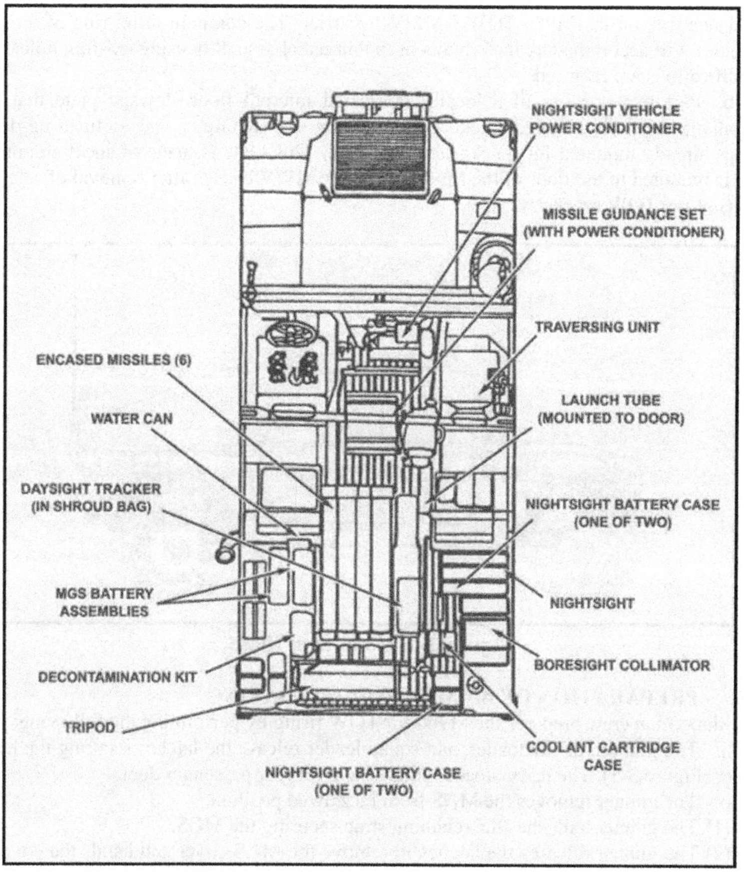

Figure 3-3. Stowing M220A2 on M966.

3-5. HMMWV INTERCHANGEABLE MOUNT SYSTEM

The HMMWV Interchangeable Mount System (HIMS) is used to convert an M966 TOW HMMWV carrier into an automatic weapons platform (Figure 3-4, page 3-6). The HIMS provides units equipped with M966 TOW HMMWV carriers with low cost, quick, and effective options for tailoring forces for contingency missions. The HIMS contains two parts. (The HIMS technical data package contains the information needed to operate the HIMS and where to find the materials needed. TM 9-2320-280-10 contains the combat loading plans for the vehicle.)

a. Part one is an armament mount panel assembly with pintle that mounts on the M1025 armament HMMWV carrier. This part is interchangeable with the missile guidance tray on the M966 TOW HMMWV carrier. The common turret ring of the two vehicles will accommodate the armament mount panel assembly using existing holes. No modifications are required.

b. Part two consists of a locally fabricated internal floor stowage plate that has automatic weapons stowage brackets, ammunition stowage trays, and footman tie-down straps already mounted for quicker emplacement. This plate is made of sheet aluminum and is mounted in the floor of the M966 TOW HMMWV carrier after removal of selected items of the TOW bracketry.

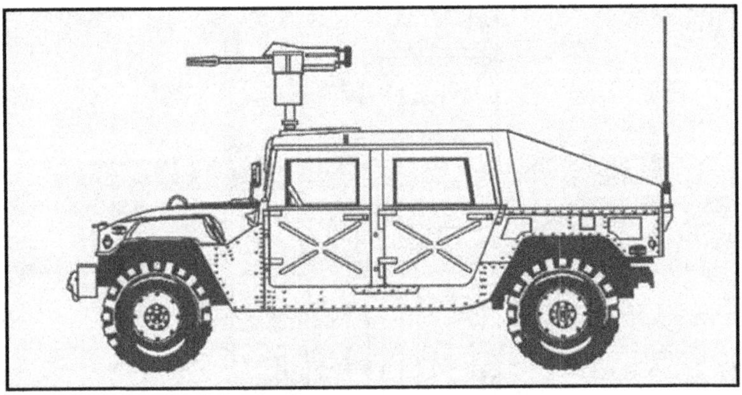

Figure 3-4. M966 with HIMS.

3-6. PREPARATION OF AN M966 FOR TOW FIRING

The three-man crew prepares the M966 for TOW firing by performing the following:

a. The gunner, driver-loader, and squad leader release the latches securing the hatch cover (Figure 3-5). The driver-loader enters the right rear passenger door.

b. The gunner removes the MGS from its stowed position.

(1) The gunner unlatches the retaining strap securing the MGS.

(2) The gunner releases the latches to remove the MGS cover and hands the cover to the squad leader.

(3) The squad leader stows the MGS cover in the front passenger's footwell of the vehicle.

(4) The gunner tilts the MGS to the vertical position on the gunner's platform and releases the lower coupling clamp on the traversing unit adapter.

c. The gunner positions the weapon station. The gunner opens the hatch and enters the weapon station, secures the hatch with the retaining latch, and rotates the weapon station by pulling up on the brake handle so that the backrest is toward the rear of the vehicle. To lock the weapon station, the gunner pushes down on the brake handle.

d. The gunner lifts the MGS through the weapon station, places it on the MGS tray, and secures it with the straps (Figure 3-6, page 3-8).

e. The gunner installs the traversing unit.

(1) The gunner opens the pedestal mount cover and secures it (Figure 3-7, page 3-8).

(2) The gunner lifts the traversing unit with adapter attached from the stowage base and secures the traversing unit to the pedestal mount with the coupling clamp ensuring that the traversing unit azimuth lock is toward the rear of the vehicle.

(3) The squad leader pulls the coil cable out of its retaining cup and hands it to the gunner (Figure 3-8, page 3-9).

f. The gunner connects the coil cable connector by aligning the yellow indexing line on the coil cable connector with the yellow indexing line on the J1 connector on the MGS (Figure 3-9, page 3-9). He turns the nut on the coil cable connector in the direction of the arrow until the red mark (band) on the M1 connector on the MGS cannot be seen. (Do not apply downward pressure on the coil cable connector. This can cause misalignment and bent pins.)

CAUTION
Ensure that no dirt is in the end of the coil cable connector. Dirt can cause a bad connection with the MGS, resulting in poor system operations.

NOTE: On the MGS for the basic TOW, push down to seat the coil cable connector on the M1 connector and tighten the locking nut.

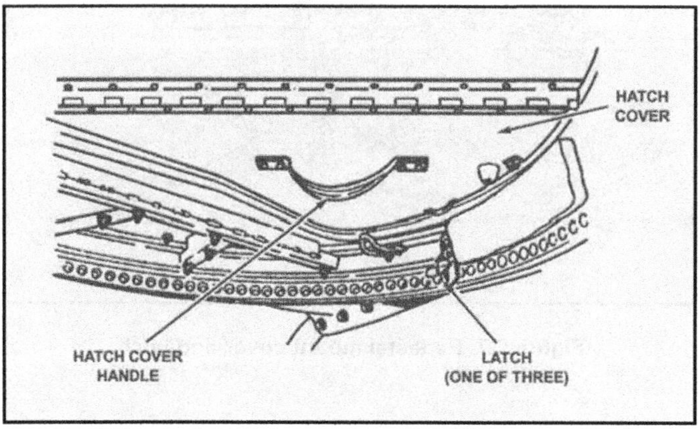

Figure 3-5. Hatch cover and latch.

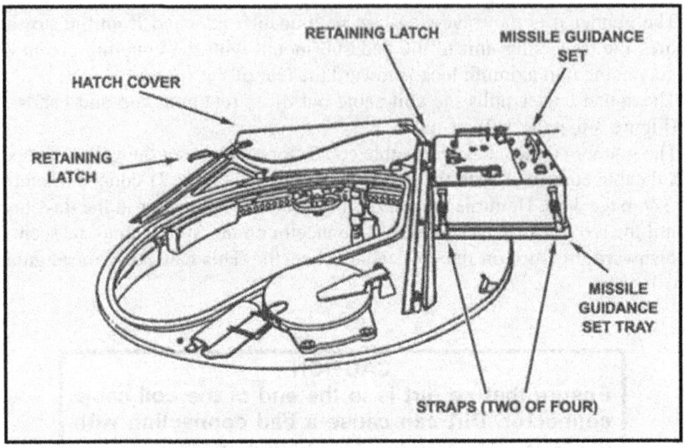

Figure 3-6. Hatch cover retaining latches and MIGS tray.

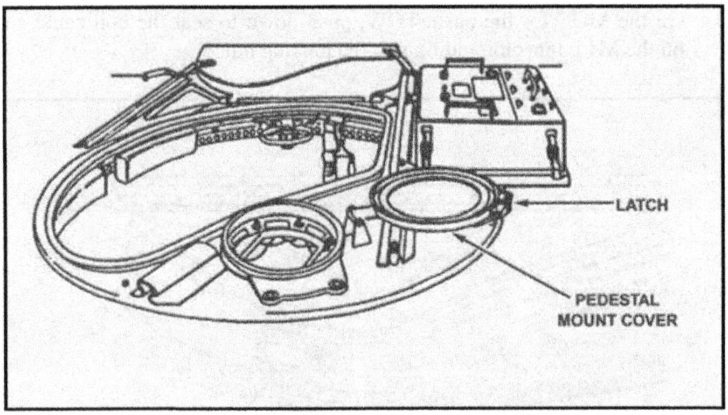

Figure 3-7. Pedestal mount cover and latch.

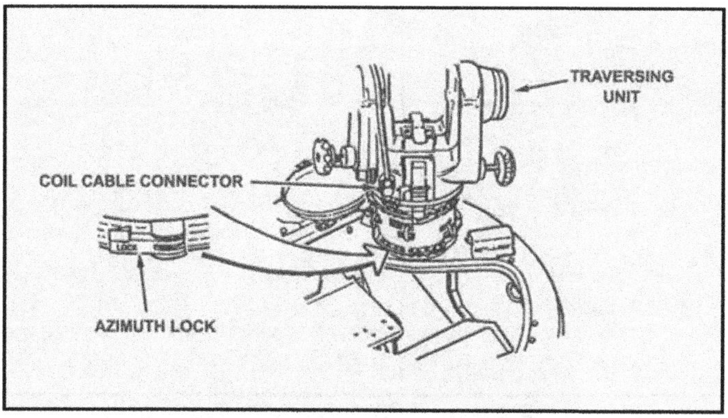

Figure 3-8. Coil cable connector.

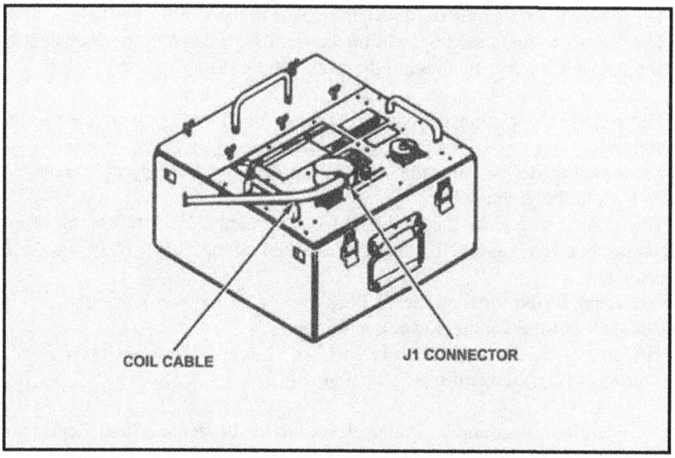

Figure 3-9. Coil cable connector and J1 connector.

g. The gunner installs the launch tube.

(1) The driver-loader uses the forward latch to open the cargo shell door and hands the launch tube to the gunner.

(2) The gunner installs the launch tube on the traversing unit and secures it with the launch tube latch (Figure 3-10, page 3-10).

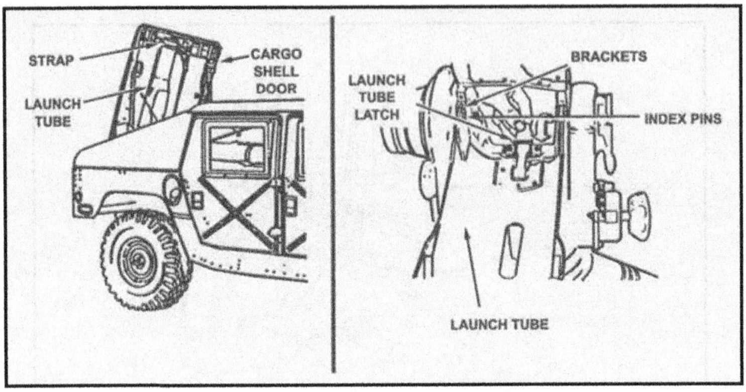

Figure 3-10. Launch tube latch.

h. The gunner installs the daysight tracker.

(1) The driver-loader hands the daysight tracker to the gunner.

(2) The gunner mounts the daysight tracker on the traversing unit boresight plate. He secures the daysight tracker by locking the latch handle (Figure 3-11).

(3) The driver-loader stows the nightsight case in its bracket.

(4) The squad leader connects connector P2 of the TVPC power cable to the power conditioner cable.

(5) The squad leader unwinds the nightsight power conditioner cable and TVPC cable and hands both to the gunner.

(6) The gunner connects the nightsight power cable 2W2 to the nightsight input power conditioner J1 (Figure 3-12, page 3-12), then connects the TVPC power cable P1 to connector J1.

(7) The squad leader turns on the TVPC.

(8) The gunner turns on the power conditioner.

(9) The gunner checks steps 10 through 15 of system self-test on TOW 2 (positions 1 and 7 on basic TOW), then collimates the nightsight.

NOTE: The limiter assembly limits elevation to 20 degrees and depression to 10 degrees. If the limiter is used, the gunner makes sure both elevation and depression limiters are in the up position and pinned (Figure 3-13, page 3-12).

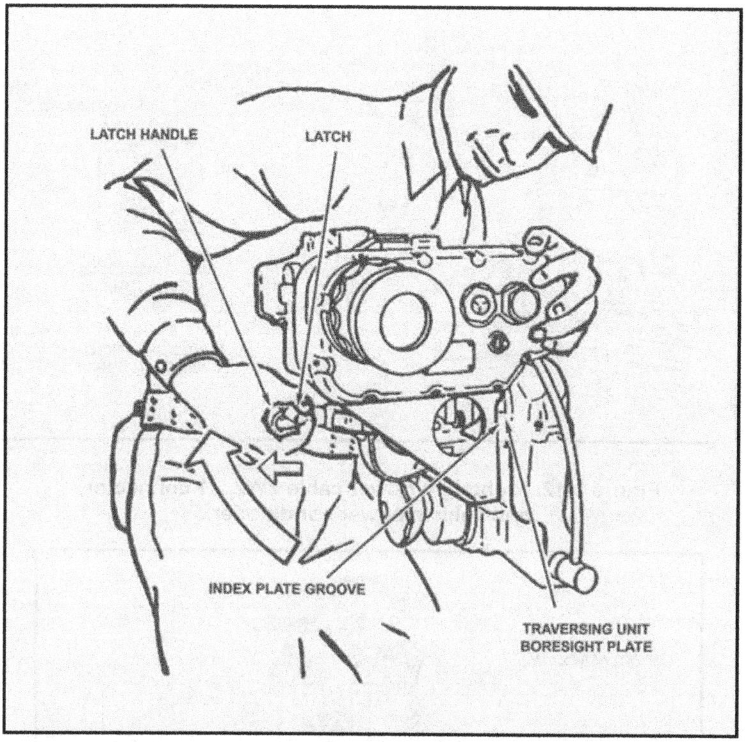

Figure 3-11. Daysight tracker on traversing unit boresight plate.

> **WARNING**
>
> **Injury to the gunner and damage to the equipment can occur if the TOW system is fired at angles greater than 20 degrees or less than -20 degrees for all firing positions except over the cargo shell door. If firing over the cargo shell door, firing angles should not be greater than 20 degrees or less than -13 degrees. Pressure waves and flying debris can cause injury to personnel and damage to equipment.**

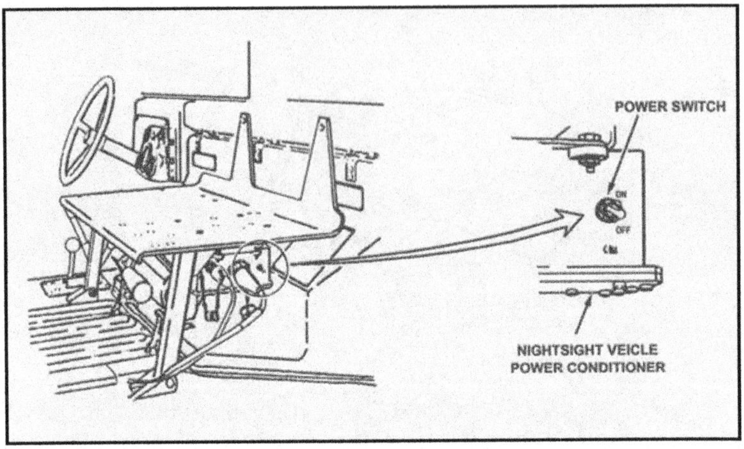

Figure 3-12. Nightsight power cable 2W2, J1 connector, and vehicle power conditioner.

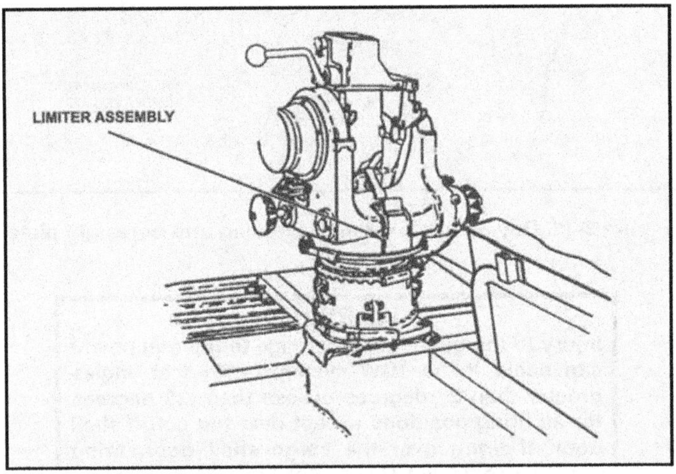

Figure 3-13. Limiter assembly location.

3-7. LOADING, ARMING, AND UNLOADING

The procedures for loading, arming, and unloading the M966-mounted TOW/TOW 2 are as follows:

a. **Load.** Missiles must be unloaded from the missile racks in numerical order. The forward handling ring, preformed packing, and quick-release clamp are saved in case the missile is not fired. If missile diaphragms get damaged while loading, the missile can still be fired. The following procedures are performed to load the M966-mounted TOW/TOW 2.

WARNING
Never open one end of the cargo shell door until the opposite end is securely closed. Personnel injury, equipment damage, or mission abort will occur if both ends open at the same time.

(1) On the traversing unit, the gunner ensures that the launch tube is locked in the 8-degree down position. He pushes the locking handle forward, then up, and opens the bridge clamp.

CAUTIONS

1. Do not drop an encased missile. Handling rings or launch container may be bent or damaged.
2. Do not break the diaphragm on the end of an encased missile. If water gets into the launcher container, the missile can be damaged.
3. Do not dispose of bad missiles. If an encased missile is damaged or handling rings are badly bent, return the encased missile to the ammunition unit for inspection.
4. Be careful not to punch a hole in an encased missile diaphragm when the forward handling ring is removed.

(2) The driver-loader unlatches and raises the cargo shell door using the forward latch. He removes the quick-release clamp and forward handling ring from the encased missile.

WARNING
Clear all personnel from the firing danger zone before proceeding with missile loading procedure.

(a) The driver-loader turns the inside portion of the protective cover two complete turns counterclockwise, and he removes it from the electrical connector. (The protective cover is kept for future use.)

(b) He turns the encased missile so that the electrical connector is facing up and hands it to the gunner.

(3) The gunner takes the encased missile from the driver-loader. He raises the back end of the missile and slides it forward and down into the launch tube until the missile indexing lugs are firmly in place. The gunner lowers the back end of the encased missile ensuring that the electrical connector joins with the bridge clamp. He lowers the bridge clamp and pushes down on the top of it. He pulls the bridge clamp locking handle downward and backward to lock the encased missile in the launch tube.

DANGER

CARGO SHELL DOOR MUST BE CLOSED AND SECURELY LATCHED AT BOTH ENDS BEFORE FIRING A MISSILE. FIRING A MISSILE WITH THE DOOR NOT SECURE WILL CAUSE INJURY TO PERSONNEL, DAMAGE TO EQUIPMENT, OR MISSION ABORT.

b. **Arm.** The following procedures are performed to arm the M966-mounted TOW/TOW 2.

(1) The driver-loader closes the cargo shell door.

(2) The gunner makes sure the cargo shell door is securely latched at the forward end.

WARNING

Do not raise arming lever until ready to fire at selected target.

(3) The gunner checks the backblast area before raising the arming lever.

CAUTION

Do not press the trigger with an empty missile case in the launch tube. Possible shorts can result, damaging the prefire, fire, and wire-cutter circuits.

c. **Unload.** The gunner lowers the trigger cover over the trigger after the missile has impacted and locks the traversing unit in the 8-degree down position. If a missile is prepared for firing and then not fired, it must be immediately tagged to show the time, date, and using unit. If the missile is not used within 90 days of initial preparation, it must be turned in to the ammunition supply point.

CAUTION
If the missile is not fired, lower the arming lever before opening the bridge clamp. The command-link wires will be cut if the arming lever is not lowered.

3-8. TARGET ENGAGEMENT

The crew members perform the following target engagement procedures.

a. When the squad leader issues the fire command, the gunner releases the azimuth and elevation locks on the traversing unit and turns on the nightsight if necessary. If crew members other than the gunner are in or leaning on the M966, they must be very still while the gunner is tracking. Even small movements can cause the vehicle to move and adversely affect the gunner's aim.

b. The gunner acquires and begins to track the target.

(1) If the target cannot be acquired in the sight, he announces, "Lost."

(2) When the target has been acquired, he identifies it as friendly or enemy.

(3) If the target cannot be identified, he announces, "Cannot identify."

(4) If the target is identified as enemy, he places the crosshairs on the center mass of the exposed portion of the target, presses the trigger, and smoothly tracks the target until missile impact.

(5) If the target is sighted out of the sector of fire, the gunner must rotate the weapon station to engage the target after the squad leader verifies the target's engageability.

3-9. IMMEDIATE ACTION

The crew members perform the following immediate action procedures in the event of a hangfire or misfire. If the gunner presses the trigger and nothing happens after 1.5 seconds, a misfire has occurred. If he presses the trigger and the pop of the explosive squib on the gyroscope is heard but the missile does not launch after 1.5 seconds, a hangfire has occurred.

DANGER
THE CREW MUST USE EXTREME CARE TO AVOID INJURY FROM A DELAYED MISSILE LAUNCH. A DELAYED LAUNCH USUALLY OCCURS WITHIN 1 MINUTE. HOWEVER, IT COULD OCCUR AS LONG AS 30 MINUTES AFTER PRESSING THE TRIGGER. KEEP ALL PERSONNEL OUT OF FIRING DANGER ZONES.

DANGER

DO NOT OPEN THE CARGO SHELL DOOR DURING FAILURE TO LAUNCH PROCEDURES. SHOULD THE MISSILE FIRE WITH THE CARGO SHELL DOOR OPEN, SEVERE INJURY TO PERSONNEL OR DAMAGE TO EQUIPMENT WILL RESULT.

a. For a *hangfire*, the gunner alerts the crew by announcing, "Hangfire" and continues to track the target for one minute.

(1) If the missile fails to launch after one minute, he closes the trigger cover, lowers the arming lever and locks the elevation lock and the azimuth lock, ensuring that the launch tube is locked in the 8-degree down position (Figure 3-14).

NOTE: In a training situation, the crew dismounts the vehicle, closes the doors, calls EOD, and waits 30 minutes. The gunner may then remount the vehicle.

DANGER

IF THE VEHICLE IS NOT POSITIONED WITH ITS FRONT OR REAR IN LINE WITH THE TARGET, IT MUST BE REPOSITIONED SO THAT EITHER THE FRONT OR REAR OF THE VEHICLE IS IN LINE WITH THE TARGET. THE WEAPON STATION MUST BE ROTATED DURING VEHICLE REPOSITIONING SO THE FORWARD END OF THE MISSILE IS POINTING TOWARD THE TARGET AT ALL TIMES.

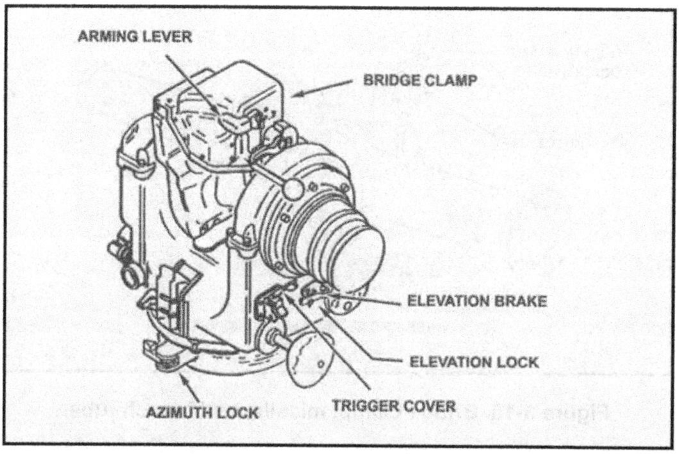

Figure 3-14. Trigger cover, arming lever, elevation lock, and azimuth lock.

(2) He opens the bridge clamp (Figure 3-15, page 3-18), removes the missile from the launch tube (keeping the missile pointed downrange at all times), and hands the missile over either side of the vehicle to the driver-loader (Figure 3-16, page 3-18).

WARNING

Do not hand off a malfunctioned missile to the front or rear of the vehicle. A malfunctioned missile must be handed off from the vehicle to either the left or right side.

(3) The driver-loader, keeping the missile pointed downrange, places the missile on the ground a safe distance from the vehicle. (In training, he places the missile on the ground at least 200 meters from the vehicle and marks the location.)

b. For a *misfire,* the gunner continues to track the target for one minute and presses the trigger again. He alerts the crew by announcing, "Misfire" and continues to track the target.

(1) If the missile still fails to launch, he disarms the missile by closing the trigger cover, lowers the arming lever, and locks the azimuth and elevation locks ensuring that the launch tube is locked in the 8-degree down position.

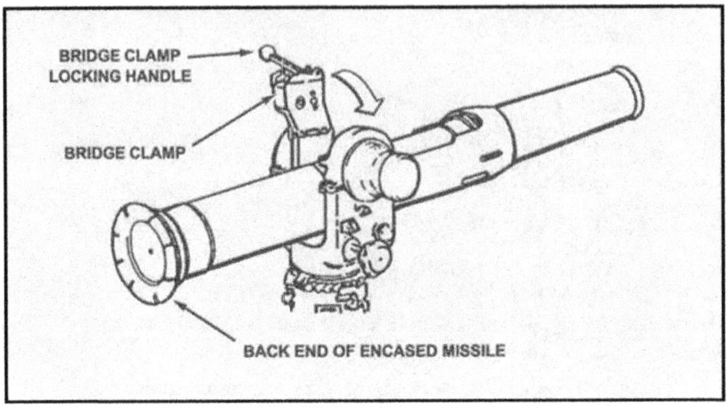

Figure 3-15. Bridge clamp, missile, and launch tube.

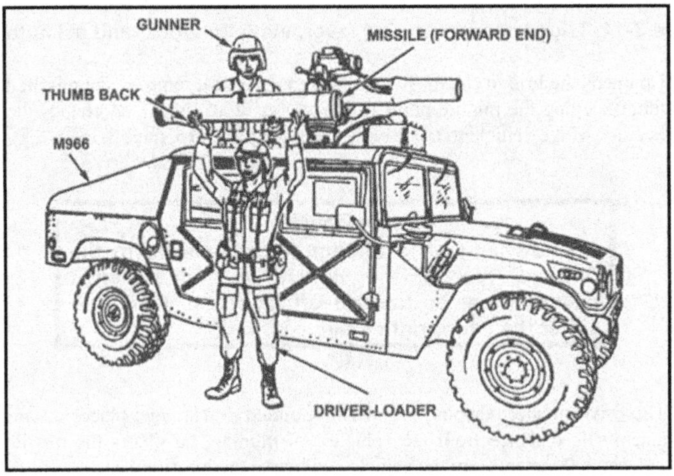

Figure 3-16. Driver-loader's position for receiving the missile from the gunner.

(a) The gunner presses the TEST-OPERATE switch to check the battery power. If he receives an OUT-OF-BAND/BATTERY FAIL, he checks to ensure that the TVPC switch is in the ON position and that the P1 cable is securely mated to the container. If all cables are connected securely, he replaces the power conditioner with a battery and checks the battery (Figure 3-17).

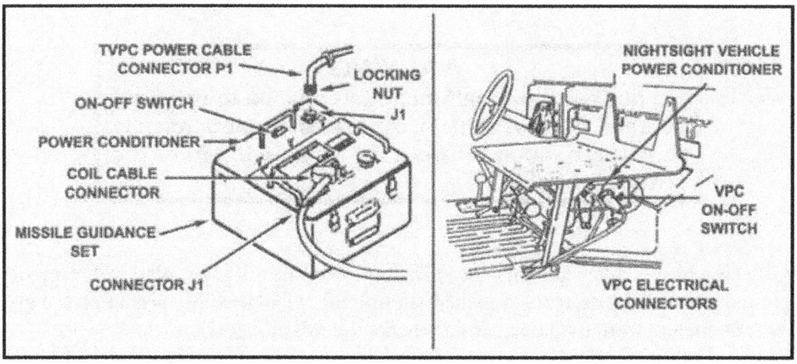

Figure 3-17. MGS, TVPC cable connector P1, power conditioner.

(b) He checks the coil cable connection to J1 connector. If it is not connected properly, he corrects it.

(c) He ensures the bridge clamp locking handle is properly seated, unlocks the azimuth and elevation locks, and raises the arming lever.

(d) He ensures the electrical connector is properly mated, raises the trigger cover, and presses the trigger.

(2) If the missile still fails to launch, the gunner may continue tracking the target for one minute (at the gunner's discretion in combat).

(a) He closes the trigger cover, lowers the arming lever, and locks the azimuth and elevation locks, ensuring the launch tube is locked in the 8-degree down position.

NOTE: In a training situation, the crew dismounts the vehicle, closes the doors, calls EOD, and waits 30 minutes. The gunner may then remount the vehicle.

DANGER

IF THE VEHICLE IS NOT POSITIONED WITH ITS FRONT OR REAR IN LINE WITH THE TARGET, REPOSITION IT SO THAT EITHER THE FRONT OR REAR OF THE VEHICLE IS IN LINE WITH THE TARGET. ROTATE THE WEAPONS STATION DURING VEHICLE REPOSITIONING SO THE FORWARD END OF THE MISSILE IS POINTING TOWARD THE TARGET AT ALL TIMES.

(b) The gunner opens the bridge clamp, removes the missile from the launch tube (keeping the missile pointed downrange at all times), and hands the missile over either side of the vehicle to the driver-loader.

WARNING
Do not hand off malfunctioned missile to the front or rear of the vehicle. A malfunctioned missile must be handed off from the vehicle to either the left or right side.

(3) The driver-loader, keeping the missile pointed downrange, places the missile on the ground a safe distance from launcher. (In training, he places the missile on the ground at least 200 meters from the launcher and marks the missile location.)

3-10. M220A1 DISMOUNT

Specific actions must be followed when dismounting the M220A1 and moving it to the firing site. All actions happen simultaneously.

a. **Dismount Actions.** Each crew member begins executing his assigned responsibilities as soon as the squad leader commands, DISMOUNT THE TOW.

(1) ***Squad Leader's Actions***. The squad leader—

(a) Turns the VPC to the OFF position, places the MGS cover on the driver-loader's seat, and places the coil cable in its retaining cup on the traversing unit.

(b) Dismounts and goes to the rear of the vehicle; opens the cargo shell door using the rear latch; lowers the tailgate; removes the tripod, battery case, and coolant cartridge case and places them on the ground beside the launch tube (on left front side of vehicle in front of the MGS and the daysight tracker); closes the tailgate and cargo shell door.

(c) Enters the right rear passenger door, gets the collimator case and the nightsight case, dismounts the vehicle, and places the collimator case and the nightsight case on the ground beside the vehicle.

(d) Carries the tripod, the coolant cartridge case, the battery case, and the launch tube to the firing site.

(2) ***Gunner's Actions*** (assisted by driver-loader when needed). The gunner—

(a) Disconnects the coil cable from the MGS; disconnects the VPC and TVPC cables.

(b) Removes the nightsight and hands it to the driver-loader who places it on the ground. Removes the daysight tracker and hands it to the driver-loader who places it on the ground in front of the left side of the vehicle. (The nightsight should be put into its case and placed behind daysight tracker.)

(c) Removes the launch tube and hands it to the driver-loader who places it on the ground in front of the daysight tracker.

(d) Unlocks the top coupling clamp on the traversing unit adapter; removes the traversing unit (minus adapter); hands the traversing unit to the driver who places it on the ground behind the daysight tracker.

(e) Removes the TVPC from the MGS and places it to his rear on top of the vehicle within reach of the cargo hatch (see next step); unstraps the MGS and hands it to the driver-loader who places it on the ground beside the daysight tracker.

(f) Opens the cargo shell door by using the forward latch; switches the TVPC for the TOW battery assembly, and hands the battery assembly to the driver-loader to install in the MGS.

(g) Unstraps a missile and hands it to the driver-loader who places the missile on the ground behind the traversing unit. The gunner then closes the cargo shell door and dismounts the vehicle.

(h) Carries the daysight tracker and the MGS to the firing site.

(3) ***Driver-Loader's Actions***. The driver-loader—

(a) Dismounts the vehicle and waits to receive equipment from the gunner.

(b) Aligns equipment as described in paragraphs 2a through 2g; installs the battery assembly in the MGS and then replaces the MGS cover on the MGS.

(c) Carries the traversing unit on the first trip, the missile on the second trip, and the nightsight and boresight collimator on the third trip to the firing site.

b. **Firing Position Actions.** Upon reaching the firing position, the crew proceeds with the following actions.

(1) ***Squad Leader's Actions.*** The squad leader—

(a) Places the coolant cartridge case, battery case, and launch tube on the ground. Sets up the tripod with the coupling clamp handle facing the direction of fire and unlocked.

(b) Pulls the coil cable out of the traversing unit retaining cup, installs the launch tube, and observes downrange for enemy targets.

(c) Issues fire commands.

(2) ***Gunner's Actions***. The gunner—

(a) Places the MGS on the ground next to the tripod and installs the daysight tracker on the traversing unit.

(b) Removes the MGS cover and connects the coil cable to the MGS.

(c) Installs the nightsight and turns the actuator switch to the ON position.

(d) Conducts the system self-test, boresight, and collimation.

(e) Engages the target upon receipt of the fire command.

(3) ***Driver-Loader's Actions***. The driver-loader—

(a) Installs the traversing unit on the tripod.

(b) Returns to the vehicle for the missile; places the missile on the ground at the firing position to the right side of the weapon system.

(c) Returns to the vehicle to get the nightsight case and the boresight collimator case; returns to the firing position and hands the nightsight to the gunner.

(d) Places the collimator on the ground, removes it from its case, and installs it on the nightsight. Attaches the battery to the collimator.

(e) Removes the collimator from the nightsight; removes the battery from the collimator; places both the collimator and the battery back in their cases (if time permits).

(f) Loads the missile and arms it if commanded.

3-11. M220A1 REMOUNT

The following actions occur at the same time, when the squad leader commands, CEASE TRACKING, OUT OF ACTION.

a. **Firing Position Actions.** The following actions take place at the firing position before returning to the vehicle.

(1) ***Squad Leader's Actions***. The squad leader—

(a) Removes the launch tube and places it on the ground.

(b) Folds the tripod legs and commands MOVE OUT. (The squad leader remains at the firing position until the driver-loader returns and all equipment is en route to the vehicle.)

(c) Carries the coolant cartridge case, tripod, battery case, and launch tube back to the vehicle. Places the launch tube on the ground on the driver's side of the vehicle and the rest of the equipment on the ground behind the vehicle.

(2) ***Gunner's Actions***. The gunner—

(a) Disconnects the coil cable.

(b) Replaces the nightsight lens cover, removes the nightsight, and places it on the ground near the driver-loader.

(c) Removes the daysight tracker and picks up the MGS; carries them to the vehicle.

(d) Places the daysight tracker and the MGS on top of the vehicle and enters the gunner's station.

(3) ***Driver-Loader's Actions***. The driver-loader—

(a) Removes the encased missile (or empty launch container if the missile was fired).

(b) Removes the traversing unit from the tripod, carries it to the vehicle, and hands it to the gunner when the gunner reenters the gunner's station.

(c) Returns to the firing position to get the nightsight and collimator; returns to the vehicle and hands the nightsight to the gunner; sets the collimator and the nightsight case on the ground near the rear of the vehicle.

(d) Returns to the firing position to get the encased missile (if a missile was not fired) and carries it to the vehicle. Hands the missile to the gunner.

b. **Remount Actions.** The following actions are performed to remount the vehicle.

(1) ***Squad Leader's Actions***. The squad leader—

(a) Stows the coolant cartridge, battery case, tripod, nightsight case, and collimator on the vehicle.

(b) Hands the TVPC and VPC cables to the gunner. Enters the squad leader's seat in the vehicle and turns the VPC to the ON position.

(c) Places the MGS cover on the floorboard.

(d) Assists other crew members as needed.

(2) ***Gunner's Actions***. The gunner—

(a) Receives the traversing unit from the driver-loader, installs it on the traversing unit adapter, and secures it with the top coupling clamp.

(b) Installs the MGS in the MGS tray and secures it with the straps; hands the MGS cover to the squad leader; connects the coil cable.

(c) Installs the daysight tracker and nightsight, and connects the VPC.

(d) Moves into the cargo area and places the unfired missile in the missile rack (if necessary).

(e) If time permits, conducts the system self-test, boresight, and collimation; exchanges the battery assembly for the TVPC; connects the TVPC cable.

(3) ***Driver-Loader's Actions.*** The driver-loader assists the squad leader and gunner as needed. He closes the tailgate and cargo shell door at the end of the drill.

NOTE: If a missile was fired, the empty launch container is destroyed.

3-12. M220A2 DISMOUNT (READY-TO-FIRE CONFIGURATION)

Specific actions must be followed when dismounting the M220A2 and moving it to the firing site. All actions happen simultaneously.

NOTE: Infantry units using the M966 conduct many types of missions. Therefore, during dismount operations, all equipment must be properly stored before leaving the vehicle.

a. **Dismount Actions.** The following actions occur at the same time when the squad leader commands, DISMOUNT THE TOW.

(1) ***Squad Leader's Actions***. The squad leader—

(a) Turns the VPC to the OFF position, places the MGS cover on the driver-loader's seat, and places the coil cable in its retaining cup on the traversing unit.

(b) Dismounts and goes to the rear of the vehicle; opens the cargo shell door using the rear latch; lowers the tailgate; removes the tripod and the BPC and places them on the ground beside the launch tube (on the left front side of the vehicle in front of the MGS and the daysight tracker); closes the tailgate and cargo shell door.

(c) Enters the right rear passenger door, gets the collimator case and the nightsight case, dismounts the vehicle, and places the collimator case and the nightsight case on the ground beside the vehicle.

(d) Carries the tripod, BPC, and launch tube to the firing site.

(2) ***Gunner's Actions*** (assisted by driver-loader when needed). The gunner—

(a) Disconnects the coil cable from the MGS and disconnects the VPC and TVPC cables.

(b) Removes the nightsight and hands it to the driver-loader who places it on the ground. Removes the daysight tracker and hands it to driver-loader who places it on the ground in front of the left side of the vehicle. (The nightsight should be put into its case and placed behind the daysight tracker.)

(c) Removes the launch tube and hands it to driver-loader who places it on the ground in front of the daysight tracker.

(d) Unlocks the top coupling clamp on the traversing unit adapter, removes the traversing unit (minus adapter), and hands it to the driver-loader who places it on the ground behind the daysight tracker.

(e) Removes the TVPC from the MGS and places it to his rear on top of the vehicle within reach of the cargo hatch (see next step); unstraps the MGS and hands it to the driver-loader who places it on the ground beside the daysight tracker.

(f) Opens the cargo shell door by using the forward latch; exchanges the TVPC for the TOW battery assembly; hands the battery assembly to the driver-loader to install in the MGS.

(g) Unstraps a missile and hands it to the driver-loader who then places the missile on the ground behind the traversing unit. The gunner then closes the cargo shell door and dismounts the vehicle.

(h) Carries the daysight tracker and the MGS to the firing site.

(3) ***Driver-Loader's Actions***. The driver-loader—

(a) Dismounts the vehicle and waits to receive equipment from the gunner.

(b) Aligns equipment as described in paragraphs 2a through 2g; installs the battery assembly in the MGS and then installs the MGS cover on the MGS.

(c) Carries the traversing unit on the first trip, the missile on the second trip, and the nightsight and boresight collimator on the third trip to the firing site.

b. **Firing Position Actions.** Upon reaching the firing position, the crew proceeds with the following actions:

(1) ***Squad Leader's Actions***. The squad leader—

(a) Places the BPC and launch tube on the ground; sets up the tripod with the coupling clamp handle facing the direction of fire and unlocked; connects the BPC cable and turns the BPC to the ON position; places the BPC cable within reach of the gunner.

(b) Pulls the coil cable out of the traversing unit retaining cup; installs the launch tube; observes downrange for enemy targets.

(c) Issues fire commands.

(2) ***Gunner's Actions***. The gunner—

(a) Places the MGS on the ground next to the tripod and installs the daysight tracker on the traversing unit.

(b) Removes the MGS cover and connects the coil cable to the MGS.

(c) Installs the nightsight and turns the actuator switch to the ON position.

(d) Conducts the system checkout, boresight, and collimation.

(e) Engages the target upon receipt of the fire command.

(3) ***Driver-Loader's Actions***. The driver-loader—

(a) Installs the traversing unit on the tripod.

(b) Returns to the vehicle for the missile; places it on the ground at the firing position to the right side of the weapon system.

(c) Returns to the vehicle and gets the nightsight case and collimator case; returns to the firing position, takes the nightsight out of the case, and hands it to the gunner.

(d) Places the collimator case on the ground; removes the collimator from its case and installs it on the nightsight. Attaches the boresight cable to the collimator.

(e) Disconnects the boresight cable and removes the collimator from the nightsight; places the collimator in its case (time permitting).

(f) Loads the missile and arms it upon receipt of the fire command.

3-13. M220A2 REMOUNT (READY-TO-FIRE CONFIGURATION)

The following actions occur at the same time, when the squad leader commands, CEASE TRACKING, OUT OF ACTION.

a. **Firing Position Actions**. The following actions are performed at the firing position.

(1) ***Squad Leader's Actions***. The squad leader—

(a) Removes the launch tube and places it on the ground.

(b) Places the coil cable in the retaining cup in the bottom of the traversing unit.

(c) Folds the tripod legs and commands MOVE OUT. (The squad leader remains at the firing position until the driver-loader returns and all equipment is enroute to the vehicle.

(d) Carries the BPC, tripod, and launch tube back to the vehicle. Places the launch tube on the ground on the driver's side of the vehicle and the rest of the equipment on the ground behind the vehicle.

(2) ***Gunner's Actions***. The gunner—

(a) Disconnects the coil cable.

(b) Replaces the nightsight lens cover; removes the nightsight and sets it on the ground near the driver-loader.

(c) Removes the daysight tracker, picks up the MGS, and carries them to the vehicle.

(d) Places the daysight tracker and MGS on top of the vehicle and enters the gunner's station.

(3) ***Driver-Loader's Actions***. The driver-loader—

(a) Removes the encased missile (or empty launch container if the missile was fired).

(b) Removes the traversing unit from the tripod, carries it to the vehicle, and hands it to the gunner when the gunner reenters the gunner's station.

(c) Returns to the firing position to get the nightsight and collimator; returns to the vehicle and hands the nightsight to the gunner; sets the collimator and the nightsight case on the ground near the rear of the vehicle.

(d) Returns to the firing position, gets the encased missile (if the missile was not fired), and carries it to the vehicle. Hands the missile to the gunner.

b. **Remount Actions.** The following actions are performed to remount the vehicle.

(1) ***Squad Leader's Actions.*** The squad leader—

(a) Stows the BPC, tripod, collimator, and nightsight case on the vehicle.

(b) Hands the TVPC and VPC cables to the gunner. Enters the squad leader's seat in the vehicle and turns the VPC to the ON position.

(c) Places the MGS cover on the floorboard.

(d) Assists the other crew members as needed.

(2) ***Gunner's Actions.*** The gunner—

(a) Receives the traversing unit from the driver-loader, installs it on the traversing unit adapter, and secures it with the top coupling clamp.

(b) Installs the MGS in the MGS tray and secures it with the straps; hands the MGS cover to the squad leader; connects the coil cable.

(c) Installs the daysight tracker and the nightsight, and connects the VPC.

(d) Moves into the cargo area and places the unfired missile in the missile rack (if necessary).

(e) Conducts the system checkout, boresight, and collimation (if time permits); exchanges the battery assembly for the TVPC and connects the TVPC cable.

(3) ***Driver-Loader's Actions.*** The driver-loader assists the squad leader and gunner as needed.

NOTE: If a missile was fired, the empty launch container is destroyed.

3-14. FIRING LIMITATIONS

Firing limitations for the M966-mounted TOW/TOW 2 are the same as for the tripod-mounted TOW/TOW 2 with an additional limitation: Restricted firing angles are those greater than +20 degrees or less than -20 for all firing positions except firing over the cargo shell door. The firing angle over the cargo shell door should not exceed +20 degrees or -13 degrees.

CHAPTER 4
TOW TRAINING PROGRAM

The TOW Training Program is a comprehensive program of required training and testing that begins with individual Skill Level 1 tasks and progresses through individual gunner qualification to squad-, section-, and platoon-level testing and verification. This process has been condensed and unified into a program that consists of 12 TOW gunnery tables plus a quarterly Gunner's Skill Test. All mandatory TOW training and testing are included in this program. Additional types of TOW training, such as STXs and CO EXEVALs, will be performed with the frequency prescribed in STRAC and in the manner described in ARTEP 7-91. The unit commander must assess the training status of the individual unit to determine what additional TOW training is necessary to support the 12 mandatory training tables and must prepare his crews to pass their required testing.

Section I. TRAINING ASSESSMENT AND PLANNING

Differences between resources required and resources available will affect both the time required to conduct and sustain training and the unit's ability to meet the required standards. Required resources and maintenance assistance should be planned for and requested well in advance. Nothing should be left to chance, because resources will be wasted and training opportunities lost. A good program must have provisions for routine inspection of the TOW weapon systems by direct support units. Leaders must adjust the training program to meet the changing needs of the unit. (See Table 4-1, page 4-2, for ammunition requirements according to DA Pam 350-38).

4-1. COMMANDER'S RESPONSIBILITY

All mandatory TOW gunnery training and testing are included in the 12 TOW Gunnery Tables plus the quarterly Gunner's Skill Test. However, the commander must still conduct an accurate assessment of the state of TOW training in his unit to determine what additional training is necessary to prepare his unit for the quarterly and semiannual tables. *He must remember that his units must meet the standards outlined in DA PAM 350-38 to be certified in the training status C1, fully trained.* A good assessment of training requirements and thorough planning produce an effective unit training program. To accomplish this, the commander should start early, be thorough, and be flexible and creative. All aspects of the training program should be coordinated to produce the desired results.

a. The commander should refer to FM 25-101 when developing his unit training program. This FM applies the doctrine established in FM 25-100 and assists leaders in the development and execution of training programs.

b. The heart of an effective training program is the development of a METL and the soldier tasks, leader tasks, and collective tasks that support the METL. Battle focus drives the METL development process; the METL is based on the wartime mission, and the unit must train as it plans to fight. Chapter 2 of FM 25-101 illustrates this process in clear and practical terms.

EVENT/TABLE	TRC A* (M80 TOW blast simulator)	TRC B (M80 TOW blast simulator))	TRC C (M80 TOW blast simulator)	TRC D (M80 TOW blast simulator)
GST	0	0	0	
TABLES 1 & 2	0	0	0	
TABLES 3 & 4	0	0	0	
TABLES 5 & 6	240	60	0	
TABLES 7 & 8	40	20	0	
TABLES 9 & 10	48	24	0	
TABLES 11 & 12	48	24	0	

***The STRAC determines a unit's ammunition requirements based on its training readiness condition (TRC). The four major categories are: TRC A, active Army; TRC B, National Guard enhanced brigades; TRC C, USAR and other National Guard units; and TRC D, Reserve training divisions.**

Table 4-1. Required TOW gunnery tables and ammunition needed per system per year (IAW DA Pam 350-38 STRAC).

4-2. SUSTAINMENT TRAINING AND EVALUATION

As individual TOW crewmen, squads, sections and platoons become qualified, the commander should maintain that status by sustainment training and evaluation, and by crew stabilization. Personnel changes are inevitable in units. Before TOW squads lose their crew integrity and combat readiness, personnel changes should be examined in detail and alternative solutions sought where possible. Figure 4-1 is an example of a crew stability management chart used to record training milestones and qualifications.

4-3. TOW TRAINING GUIDELINES

A soldier must track all targets and perform all individual and collective tasks and battle drills discussed in this manual to become a fully trained heavy antiarmor weapon system crewman. Soldier training publications and ARTEP battle drills for TOW provide tasks, conditions, and standards for combat-critical skills.

a. When a crew member becomes skilled in his job, he should be trained to do another crew member's job (cross-trained). Therefore, the loss of one crew member will not make the squad combat-ineffective.

b. Training should be conducted under realistic conditions. To properly develop combat skills, training must be conducted in a tactical environment with emphasis on the type of threat the crew can expect to face in combat.

c. Multiechelon training should be conducted to save time and resources. For example, while squads are training in crew tasks, leaders can train to execute their tactical command and control responsibilities.

d. Training should be conducted with supported units. When possible, tactical training should be conducted with the type of units to be supported in combat—infantry, mechanized infantry, and armor. TOW units must be able to interface with the units they support.

VEH NO.	TOW DUTY POSITION	JAN 93	FEB 93	MAR 93	APR 93	MAY 93	JUN 93
A-31	SECTION LEADER	8278 (NOTE 2)	8278	8632 (NOTE 2)	8632	8632	8632
A-31	GUNNER	7257	7257	7257 GST (NOTE 3)	7257	7257	7257
A-31	LOADER	4410	4410	4410 GST	4410	4410	4410
A-31	DRIVER	3981	3981 GST	1850 SQD QUAL	1850	1850	1850
A-32	SQUAD LEADER	8273	8273 GST	1711 SQD QUAL	1711	1711	1711
A-32	GUNNER	1984	1984 GST	1984 SQD QUAL	1984	1984	1984
A-32	LOADER	2029	2029 GST	2029 SQD QUAL	2029	2029	2029
A-32	DRIVER	9699	9699 GST	9699 SQD QUAL	9699	9699	9699

NOTES:

1. This number represents the last four of the individual's social security number. It is written in black in this particular block and remains black until the individual leaves the duty position.
2. This number is written in red, signifying the change in personnel. April 93 number goes to black.
3. Training data is colored in green to indicate successful completion; red indicates failure.

Figure 4-1. Example of a crew stability management chart.

e. Training should be concentrated in areas where it is most needed. Time should not be wasted by training skills the unit and soldiers have already mastered. The trainer must know where the weak spots are and train to correct them.

f. Sections and squads will not always be at full strength. The mission in combat and in training can still be accomplished if under-strength units are organized with the following rules in mind:

(1) Key leadership positions should always be filled. For example, the gunner may have to fill the position of the squad leader.

(2) The primary weapon system should always be manned. If the loader becomes a casualty, the squad leader may have to load the TOW and control the squad at the same time. If the gunner becomes a casualty, the squad leader must operate and fire the TOW.

4-4. TOW GUNNERY TABLES

The TOW Training Program consists of 12 gunnery tables plus the Gunner's Skill Test. The STRAC establishes the frequency of a unit's training requirements based on its TRC. A TRC A (active Army) unit, for example, is required to conduct Tables 1 through 6 and the Gunner's Skill Test and Tables 7 through 12 semiannually. (See Table 4-2 for frequency of events as required by DA Pam 350-38 STRAC.)

NOTE: The second quarterly GST should be conducted within seven days before conducting Tables 7 through 12.

EVENT/TABLE	TRC A*	TRC B	TRC C	TRC D
Weapons Training	12	4	1	1
Gunner's Sustainment	12	4	1	1
GST	4	1	1	1
TABLES 1 & 2*	4	2	1	1
TABLES 3 & 4	4	2	1	1
TABLE 5	4	1	1	1
TABLE 6	4	1	1	0
TABLES 7 & 8	2	1	0	0
TABLES 9 & 10	2	1	0	0
TABLES 11 & 12	2	1	0	0
*See STRAC Table 5-11 for details on personnel required to conduct training and any training devices used.				

Table 4-2. Frequency of required TOW gunnery events.

Tables 1 and 2 will be performed by all of the crew with the TOW Gunnery Trainer (TGT). Only gunners will perform tables 3 and 4. Tables 5 through 12 will be performed with the TOW Field Tactical Trainer (TFTT). The TFTT replicates the flight and performance characteristics of the TOW. (See Appendix F for more information on TOW training devices.)

4-5. CONDUCT OF UNIT TOW TRAINING

TOW training involves individual and collective training divided into four phases. Each unit must complete these phases as required.

a. **Phase I—Individual Training.**

- Conduct refresher training on STP tasks as deemed necessary to prepare for the Gunner's Skill Test.
- Conduct sustainment training with the TGT as deemed necessary to prepare for TOW Gunnery Tables 1 through 4.
- Conduct Gunner's Skill Test.
- Conduct TOW Gunnery Table 1, Individual Gunnery Practice.

- Conduct TOW Gunnery Table 2, Individual Gunnery Qualification.
- Conduct TOW Gunnery Table 3, Advanced Gunnery Practice.
- Conduct TOW Gunnery Table 4, Advanced Gunnery Qualification.

b. **Phase II—Squad Training.**

- Conduct refresher training on the installation and operation of TFTT or MILES equipment as deemed necessary in preparation for TOW Gunnery Tables 5 and 6.
- Conduct TOW Gunnery Table 5, Baseline Gunnery Practice.
- Conduct TOW Gunnery Table 6, Baseline Gunnery Qualification.
- Conduct squad-level sustainment training as deemed necessary to prepare for TOW Gunnery Tables 7 and 8.
- Conduct TOW Gunnery Table 7, Squad Gunnery Practice, in preparation for Table 8.
- Conduct TOW Gunnery Table 8, Squad Gunnery Qualification.

c. **Phase III—Section Training.**

- Conduct section-level sustainment training as deemed necessary to prepare for TOW Gunnery Tables 9 and 10.
- Conduct TOW Gunnery Table 9, Section Gunnery Practice, in preparation for Table 10.
- Conduct TOW Gunnery Table 10, Section Gunnery Qualification.

d. **Phase IV—Platoon Training.**

- Conduct platoon-level sustainment training as deemed necessary to prepare for TOW Gunnery Tables 11 and 12.
- Conduct TOW Gunnery Table 11, Platoon Gunnery Practice, in preparation for Table 12.
- Conduct TOW Gunnery Table 12, Platoon Gunnery Qualification.

4-6. COLLECTIVE TRAINING

Collective training is the coordinated performance of individual skills and tasks as a team. Collective tasks are normally performed by more than one soldier.

a. Common crew task training is the first level of collective training. Common crew tasks provide the basic technical collective tasks performed by the crew to accomplish the wartime mission. These tasks tell "how to" and allow each soldier in the crew to perform his tasks. As a result, the squad leader, gunner, driver, and loader combine their learned individual technical skills to perform as one crew on their assigned vehicle.

b. The critical squad and platoon collective tasks are called battle drills. Battle drills are mostly independent of METT-TC and require minimal leader actions to execute. They are standardized throughout the U.S. Army. Battle drills are in the applicable drill book for each echelon and equipment-type unit. The battle drills for antiarmor platoons are in ARTEP 7-91-Drill.

c. Common crew tasks are tasks performed by a TOW crew that are not battle drills but are performed in the same manner. An example of a common crew task is Dismount/Remount the TOW. Crew tasks are crucial to applying tactics to enhance survival and to accomplish the TOW crew's mission of providing antiarmor direct fire support to destroy the enemy.

d. ARTEP training exercises are used to practice, evaluate, and sustain collective tasks and mission proficiency. They have specific goals and are modified based on METT-T factors. Exercises are structured to the availability of resources, time, and training areas. (See appropriate ARTEP for details on conducting ARTEP exercises.)

SECTION II. TOW TRAINING PHASES

The TOW training program is divided into four phases: individual training, squad training, section training, and platoon training. Phase I (individual training) consists of the Gunner's Skill Test and TOW Gunnery Tables 1 through 4. Phase II (squad training) consists of TOW Gunnery Tables 5 through 8. Phase III (section training) consists of TOW Gunnery Tables 9 and 10. Phase IV (platoon training) consists of TOW Gunnery Tables 11 and 12. (All scoring forms are located in the back of this manual for local reproduction.)

4-7. PHASE I: INDIVIDUAL TRAINING

Individual training covers all tasks that are performed by a single TOW crewman. These tasks include Skill Level 1 11B soldier's manual tasks and individual gunnery. They are the building blocks that form the foundation of TOW training.

a. **Gunner's Skill Test.** The Gunner's Skill Test comprises the Skill Level 1 tasks for the TOW weapon system listed in STP 7-11B1-SM-TG plus some additional tasks (Figure 4-2, page 4-7 and Figure 4-3, page 4-8). All soldiers assigned to TOW-equipped units must take this portion of the test and receive a GO on at least 70 percent of the tasks. Soldiers must be tested on the group of tasks that applies to the vehicle his unit is equipped with and receive a GO on at least 70 percent of those tasks. Some of the tasks are combined individual/crew tasks. Since every member of the crew must be cross trained, the combined individual/crew tasks will be performed in rotation and every member of the crew will be tested on each part.

TASK: Perform the TOW gunner's tasks listed in (1) below.
CONDITIONS: Testing will be hands on, with the appropriate equipment. Specific conditions will be found in the reference cited for each task.
STANDARDS: Specific standards for each task listed will be found in the appropriate reference. The overall minimum standard to receive a GO on the Gunners' Skill Test is for each individual to receive a GO on at least 70 percent of the tasks he is required to test on.

---

(1) The GST consists of the following tasks.

(a) 071-056-0004 Maintain an M220-series launcher system.

(b) 071-056-0007 Load an M220-series launcher system.

(c) 071-056-0008 Unload an M220-series launcher system.

(d) 071-056-0009 Engage targets with an M220-series launcher system.

(e) 071-056-0010 Perform immediate action for an M220-series launcher system malfunction.

(f) 071-056-0005 Conduct a system check-out on an M220A1 launcher system. (Tested only if unit is equipped with M220A1 systems.)

(g) 071-056-0013 Conduct a system check-out on an M220A2 launcher system. (Tested only if unit is equipped with M220A2 systems.)

(h) 071-056-0030 Determine TOW-series launcher system firing limitations.

(i) No Task Number Identify combat vehicles. (See Appendix G.)

(j) No Task Number Determine if a target can be engaged by an M220-series launcher system. (See Chapter 5.)

(k) 071-317-0000 Prepare an antiarmor range card.

(l) No Task Number Assemble an M220-series launcher system. (See TM 9-1425-472-12 [M220A1] or TM 9-1425-450-12 [M220A2].)

GUNNER'S SKILL TEST - PART 1

For use of this form, see FM 3-22.34. The proponent agency is TRADOC.

NAME	RANK	UNIT	DATE
THOMPSON, DAVID B.	SGT	C 2/2 IN	13 AUG 03

TEST#	TASK NUMBER	TASK TITLE	GO/NO GO
1	071-056-0004	MAINTAIN AN M220-SERIES LAUNCHER SYSTEM	GO H.H.R.
2	071-056-0007	LOAD AN M220-SERIES LAUNCHER SYSTEM	GO K.L.C.
3	071-035-0008	UNLOAD AN M220-SERIES LAUNCHER SYSTEM	GO K.L.C.
4	071-056-0009	ENGAGE TARGETS WITH AN M220-SERIES LAUNCHER SYSTEM	GO K.L.C.
5	071-056-0010	PERFORM IMMEDIATE ACTION FOR AN M220-SERIES LAUNCHER SYSTEM	1. NO GO B.L.C. 2. GO B.L.C.
6	071-056-0005	CONDUCT A SYSTEM CHECK OUT ON AN M220A1 LAUNCHER SYSTEM	
7	071-056-0013	CONDUCT A SYSTEM CHECK OUT ON AN M220A2 LAUNCHER SYSTEM	GO K.M.
8	071-056-0030	DETERMINE M220-SERIES LAUNCHER SYSTEM FIRING LIMITATIONS	GO R.H.
9	NO TASK NUMBER	IDENTIFY COMBAT VEHICLES	GO A.Q.
10	NO TASK NUMBER	DETERMINE IF A TARGET CAN BE ENGAGED BY AN M220-SERIES LAUNCHER SYSTEM (SEE FM 3-22.34, CH5)	GO R.P.
11	071-317-0000	PREPARE AN ANTIARMOR RANGE CARD	GO R.P.
12	NO TASK NUMBER	ASSEMBLE AN M220-SERIES LAUNCHER SYSTEM	GO W.E.

DA FORM 7331-R, AUG 2003 APD V1.00

Figure 4-2. Example of completed DA Form 7331-R, Gunner's Skill Test—Part 1.

GUNNER'S SKILL TEST - PART 2			
For use of this form, see FM 3-22.34. The proponent agency is TRADOC.			
NAME: WILLIAMS, RAY	RANK: SPC	UNIT: CSC 2/1 IN	DATE: 13 AUG 03

TEST#	TASK NUMBER	TASK TITLE	GO/NO GO
1	NO TASK NUMBER	PLACE THE M966 VEHICLE IN THE READY-TO-FIRE CONFIGURATION (SEE FM 3-22.34 CH 3)	GO K.L.C.
2	071-056-9932	CONDUCT DISMOUNTING AND REMOUNTING OF AN M220A1 LAUNCHER SYSTEM ON AN M966 VEHICLE (M220A1- EQUIPPED UNITS ONLY)	
3	071-056-0034	CONDUCT DISMOUNTING AND REMOUNTING OF AN M220A2 LAUNCHER SYSTEM ON AN M966 VEHICLE (M220A2 - EQUIPPED UNITS ONLY)	GO K.L.C.
4	071-034-0001	LOAD AN M243 OR M259 SMOKE GRENADE LAUNCHER	
5	071-034-0002	UNLOAD AN M243 OR M259 SMOKE GRENADE LAUNCHER	

DA FORM 7332-R, AUG 2003 APD V1.00

Figure 4-3. Example of completed DA Form 7332-R, for Gunner's Skill Test—Part 2.

(4) The following are the tasks, conditions, and standards for the unnumbered tasks in the Gunner's Skill Test.

TASK: Identify combat vehicles.
CONDITIONS: Given a classroom with appropriate audiovisual equipment and assorted vehicle GTAs. (See the local Training Support Center for the combat vehicle slides.)
STANDARDS: Given a 12-second exposure time for each slide, the soldier must correctly identify 36 out of 40 vehicles by nomenclature as described in Appendix G.

TASK: Determine if a target can be engaged by an M220-series TOW launcher system.
CONDITIONS: In the field, given an operational, assembled M220-series TOW launcher system and tactical vehicles placed in realistic situations; or indoors, given a series of graphic presentations of target engageability situations or using the TGT.
STANDARDS: Correctly determine target engageability for the M220-series TOW launcher system in terms of range and obstructions using both the daysight tracker and nightsight as described in Chapter 5, paragraph 5-1.

TASK: Assemble an M220-series launcher system.
CONDITIONS: Given a complete but unassembled M220-series launcher.
STANDARDS: Correctly assemble an M220-series launcher system as described in TM 9-1425-472-12 (M220A1) or TM 9-1425-450-12 (M220A2) within 3 minutes.

TASK: Place the M966 HMMWV in the ready-to-fire configuration.
CONDITIONS: Given an M966 HMMWV, an M220-series TOW launcher complete, and an MSR.
STANDARDS: Correctly place the M966 in the ready-to-fire configuration as described in TM 9-1425-475-12 (M220A1) or TM 9-1425-450-12 (M220A2) within 5 minutes.

b. **TOW Gunnery Table 1, Individual Gunnery Practice.** TOW Gunnery Table 1 consists of firing a minimum of two practice events (instructional firing) in preparation for actual gunner qualification on TOW Gunnery Table 2. Units will use the TGT to conduct this table.

TASK: Verify on TOW Gunnery Table 1, Individual Gunnery Practice using the TGT.
CONDITIONS: Indoors, using the TGT. The gunner will fire a minimum of two 10-shot events using the TGT. They will be selected from the events listed in Table 4-3 (page 4-10).
STANDARDS: To verify on TOW Gunnery Table 1 and progress to TOW Gunnery Table 2, the gunner must fire at least one 10-shot event with a cumulative score of at least 625 points and at least 6 hits.

(1) ***Preparation of the TGT for Gunner Practice.*** The TGT is not designed for use as an outdoor trainer. Therefore, it should always be set up and used indoors. It may be used in tents or under shelter as long as it is not exposed to the elements, particularly dust or blowing sand. It may be used in either the tripod-mounted mode or in the HMMWV-mounted mode. The TGT should *never* be mounted on a moving HMMWV. It should be assembled and operated according to the instructions in TM 9-6920-452-10.

(2) ***Personnel Duties.*** Normally, only two people at a time, the gunner and the trainer, will be involved in training with the TGT. The trainer will usually be either the squad or section leader. Other members of the squad should be occupied with concurrent training. Other gunners should not be allowed to watch during conduct of the event. Once the TGT has been assembled and warmed up, the trainer must assemble a plan group that conforms to one of the 10-shot events listed in Table 4-3 (page 4-15). The trainer should have the TGT inspected and warmed up, and the plan group constructed before the gunner is brought in to begin engaging targets. When constructing the plan group, the trainer ensures battle sounds are ON, target size is set at 100 percent, and obscuration is set to last for 1.5 seconds for the thermal sight and for the daysight tracker. When everything is ready, the trainer should bring the gunner in and brief him before beginning the event. The briefing should include the following:

- Safety considerations.
- Breath control—remind the student of proper procedures.
- Obscuration—tell the student the obscuration setting used.
- Sight—tell the student which sight he will use.
- Special conditions—tell the student any special conditions (multiple targets and so forth) that apply.

(3) ***Conduct of Training.*** When the TGT is ready for operation, the plan group has been constructed, and the gunner has been briefed and is in place, the trainer begins conducting the table. Before the trainer starts each mission, he gives the gunner a fire command and tells him to fire when ready. During the mission, the trainer does not coach the gunner in any way. The gunner must determine when to fire and at which target. After each mission, the trainer gives the gunner the results and then proceeds to the next mission.

NOTE: The trainer has the option of thoroughly debriefing the gunner on his performance at the end of each mission or of saving all 10 missions and reviewing them after the table is complete. The trainer must remember that the TGT will only save a maximum of 20 missions at one time.

TABLE NO.	1	2	3
DISK NO.	DFT 10	DFT 11	DFT 12
MISSION NO.	2	1	22
	3	2	24
	6	4	26
	8	3	28
	9	18	29
	18 (THERMAL)	26 (THERMAL)	2 (THERMAL)
	14 (THERMAL)	27 (THERMAL)	3 (THERMAL)
	12 (THERMAL)	28 (THERMAL)	4 (THERMAL)
	15 (THERMAL)	29 (THERMAL)	9 (THERMAL)
	22 (THERMAL)	30 (THERMAL)	10 (THERMAL)

Table 4-3. 10-shot events for the TGT.

(4) ***Scoring.*** The two scoring criteria for TOW Gunnery Table 1 are total point score and number of hits. The point score for the table consists of the cumulative point score for all 10 missions. The minimum number of hits is 6 out of 10. The trainer should use DA Form 7333-R to record results (Figure 4-4).

(5) ***Sustainment Training.*** Units are required to conduct monthly sustainment training. This training should not be confused with the quarterly tables. Missions for monthly sustainment training may be taken from any of the videodiscs and are chosen at the trainer's discretion. The missions in Table 4-3 should not be used for sustainment training, if possible.

TOW GUNNERY TABLES 1 AND 2: INDIVIDUAL GUNNERY

[X] PRACTICE [ ] QUALIFICATION

For use of this form, see FM 3-22.34. The proponent agency is TRADOC.

DATA REQUIRED BY PRIVACY ACT OF 1974

AUTHORITY: 10USC 3012(g). Executive Order 9397. PRINCIPAL PURPOSE(S): Records individual's performance on record fire range. ROUTINE USE(S): Evaluation of individual's proficiency and basis for determination of award of proficiency badge; SSN is used for positive identification purposes only. MANDATORY OR VOLUNTARY DISCLOSURE AND EFFECT ON INDIVIDUAL NOT PROVIDING INFORMATION: Voluntary; individuals not providing information cannot be rated/scored on a mass basis.

NAME	RANK	SSN	UNIT
ERVIN, D.	SGT	000-00-0011	C 2/2 IN
DATE		DISK#	

MISSION#	INDIVIDUAL GUNNERY TASK TESTED	SIGHT	SCORE	SUCCESSFUL	
# 22		OS	90	(YES)	NO
# 24		OS	90	(YES)	NO
# 26		OS	90	(YES)	NO
# 28		OS	80	(YES)	NO
# 29		OS	85	(YES)	NO
# 2		TS	88	YES	(NO)
# 3		TS	81	(YES)	NO
# 4		TS	78	(YES)	NO
# 9		TS	66	(YES)	NO
# 10		TS	75	(YES)	NO
	TOTAL TRACKING SCORE		823	SUCCESSFUL MISSIONS: 9	

GUNNER CLASSIFICATION

EXPERT / 875 - 1,000 1ST CLASS / 750 - 874 2ND CLASS / 625 - 749

SCORING INDIVIDUAL GUNNERY TABLES

IN THE INDIVIDUAL GUNNERY TABLES, THE GUNNER'S RATING IS DETERMINED BY HIS TOTAL POINT SCORE FOR 10 MISSIONS WITH AT LEAST 6 SUCCESSFUL MISSIONS. A SUCCESSFUL MISSION IS ONE WHERE THE GUNNER CORRECTLY FOLLOWS THE TRAINERS FIRE COMMANDS AND ENGAGES AND DESTROYS THE CORRECT TARGET.

DA FORM 7333-R, AUG 2003 APD V1.00

Figure 4-4. Example of completed DA Form 7333-R, TOW Gunnery Tables 1 and 2: Individual Gunnery Practice and Qualification.

c. **TOW Gunnery Table 2, Individual Gunnery Qualification.** TOW Gunnery Table 2 will be set up and run the same as TOW Gunnery Table 1. It will be performed with the TGT.

TASK: Qualify on TOW Gunnery Table 2, Individual Gunnery Qualification using the TGT.
CONDITIONS: Indoors using the TGT. The gunner will fire a 10-shot event selected from Table 4-3.
STANDARDS: To qualify on TOW Gunnery Table 2 and progress to TOW Gunnery Table 3, the gunner must fire one 10-shot event with a cumulative score of at least 625 points and at least 6 hits. The gunner's classification (expert, first class, second class) will be determined by the cumulative point score using the point scale as shown in Figure 4-4.

NOTE: Preparation of the TGT, personnel duties, and conduct of training are the same as for TOW Gunnery Table 1.

d. **TOW Gunnery Table 3, Advanced Gunnery Practice.** TOW Gunnery Table 3 (Figure 4-5, page 4-14) consists of a series of missions performed with the TGT that train and test advanced gunnery skills. These skills include determining a target's engageability, identifying targets (to include friend or foe), prioritizing targets, deciding which fire control methods to use, and engaging evasive targets that are hard to track such as helicopters, obscured targets, or extreme-range targets. Only the assigned gunners in each squad are required to verify on TOW Gunnery Table 3. (Other members of the squad may perform the table if time permits and the trainer thinks it useful.) TOW Gunnery Table 3 can only be performed with the TGT. Gunners must verify on TOW Gunnery Table 3 to progress to TOW Gunnery Table 4, Advanced Gunnery Qualification.

(1) ***Gunner Verification on TOW Gunnery Table 3, Advanced Gunnery Practice.***

TASK: Verify on Table 3, Advanced Gunnery Practice.
CONDITIONS: Indoors, using the TGT. The gunner will fire a minimum of two 10-shot events using the TGT. The 10-shot events will be selected from Table 4-4 (page 4-15).
STANDARDS: To verify on TOW Gunnery Table 3 and progress to TOW Gunnery Table 4, the gunner must score at least 625 points. The gunner must also receive a GO on at least 6 out of 10 missions.

---

(2) ***Personnel Duties.*** The trainer (usually the squad or section leader) sets up the TGT IAW TM 9-6920-452-10, creates the planning group for each 10-shot event, briefs the gunner, gives the fire commands for each mission, and completes DA Form 7334-R (TOW Gunnery Tables 3 and 4, Advanced Gunnery Practice and Qualification Scorecard, Figure 4-5, page 4-14). He also debriefs and critiques the gunner at the end of each 10-shot event.

(3) ***Conduct of Training.*** Conduct of advanced gunnery training with the TGT is more complex than individual gunnery. The trainer must not only observe and record the gunner's tracking performance but also brief, observe, and score the particular advanced gunnery task that is presented in each separate mission. The trainer selects the 10-shot event from Table 4-4 (page 4-15). The sequence of missions in each 10-shot event should be changed continuously to prevent the gunner from becoming familiar with them.

(a) Before each mission, the trainer must present a fire command to the gunner that clearly states what the gunner must do. For example, if a particular mission involves prioritizing targets (such as tanks are higher priority targets than APCs), the trainer gives the command "GUNNER! TARGETS TO YOUR FRONT! ENGAGE TANKS FIRST! FIRE WHEN READY!"

(b) In some missions, the task involved may not require specific fire commands. These tasks include determining a target's engageability, engaging evasive/obscured targets, and identifying targets (to include friend or foe). When giving fire commands for these missions, the trainer need only give general instructions, such as "GUNNER! TARGETS TO YOUR FRONT! FIRE WHEN READY!"

(4) ***Scoring.*** The trainer has two criteria for scoring each of these missions: tracking and success of the mission. The trainer will record the gunner's performance on the DA Form 7334-R (Figure 4-5, page 4-14).

(a) *Tracking Score.* TOW Gunnery Table 3 is scored in the same manner as TOW Gunnery Table 1 except for target identification (to include friend or foe) missions. In these missions, if the gunner correctly identifies a target as friendly and does not fire on it, he receives 100 points. If he fires on a friendly target, he receives 0 points whether he hits it or not. The total tracking score from the 10 missions (Table 4-4, page 4-15) determines the gunner's ranking.

(b) *Successful Mission Score.* The gunner must achieve a GO on at least 6 out of 10 missions. The requirements for a mission to be successful vary according to the advanced gunnery task that is presented in each mission. A successful mission is defined as follows:

- The gunner successfully engages (destroys) the correct target as determined by the trainer's instructions for that particular mission. In a mission to determine a target's engageability, a hit on any target presented is scored as a success. In scenarios to prioritize targets or determine the method of fire control, the gunner must engage the correct target according to the instructions given to him in the fire command.
- In a scenario to identify targets (including friend or foe), the gunner must not fire on friendly targets.
- Engaging evasive and or obscured targets does not require special skills or actions from the gunner. These targets are simply hard to hit. A hit on any target is scored as a successful mission.

TOW GUNNERY TABLES 3 AND 4: ADVANCED GUNNERY

☐ PRACTICE ☒ QUALIFICATION

For use of this form, see FM 3-22.34. The proponent agency is TRADOC.

DATA REQUIRED BY PRIVACY ACT OF 1974

AUTHORITY: 10USC 3012g, Executive Order 9397. **PRINCIPAL PURPOSE(S):** Records individual's performance on record fire range. **ROUTINE USE(S):** Evaluation of individual's proficiency and basis for determination of award of proficiency badge; SSN is used for positive identification purposes only. **MANDATORY OR VOLUNTARY DISCLOSURE AND EFFECT ON INDIVIDUAL NOT PROVIDING INFORMATION:** Voluntary; individuals not providing information cannot be rated/scored on a mass basis.

NAME	RANK	SSN	UNIT
HEIDER, R.	SPC	000-01-0000	C 3/2 IN

DATE	DISK#
30 AUG 03	10

MISSION#	INDIVIDUAL GUNNERY TASK TESTED	SIGHT	SCORE	SUCCESSFUL	
# 1	EVASIVE TARGET	OS	81	YES (circled)	NO
# 7	TARGET ENGAGEABILITY	OS	78	YES (circled)	NO
# 12	FIRE CONTROL METHODS	TS	73	YES (circled)	NO
# 14	EVASIVE/OBSCURE TARGET	TS	66	YES (circled)	NO
# 22	TARGET PRIORITY	TS	81	YES	NO (circled)
# 24	TARGET ENGAGEABILITY	OS	83	YES (circled)	NO
# 28	EVASIVE TARGET	OS	91	YES (circled)	NO
# 29	TARGET PRIORITY	OS	78	YES (circled)	NO
# 26	TARGET PRIORITY	OS	80	YES (circled)	NO
# 13	FRATRICIDE	OS	100	YES (circled)	NO
	TOTAL TRACKING SCORE		811	SUCCESSFUL MISSIONS:	9

GUNNER CLASSIFICATION

EXPERT / 875 - 1,000 1ST CLASS / 750 - 874 2ND CLASS / 625 - 749

SCORING ADVANCE GUNNERY TABLES

IN THE INDIVIDUAL GUNNERY TABLES, THE GUNNER RATING IS DETERMINED BY HIS TOTAL POINT SCORE FOR 10 MISSIONS, PLUS HE MUST HAVE AT LEAST 6 SUCCESSFUL MISSIONS. A SUCCESSFUL MISSION IS ONE WHERE THE GUNNER CORRECTLY FOLLOWS THE TRAINERS FIRE COMMANDS AND ENGAGES AND DESTROYS THE CORRECT TARGET. IF THE MISSION IS A FRATRICIDE MISSION, THE GUNNER SHOULD **NOT** ENGAGE THE TARGET. IF HE **DOES NOT** ENGAGE THE FRIENDLY VEHICLE, HE SCORES 100 POINTS AND A SUCCESSFUL MISSION. IF HE **DOES** ENGAGE THE FRIENDLY VEHICLE, HE SCORES 0 POINTS AND AN UNSUCCESSFUL MISSION. IN TARGET ENGAGEABILITY AND EVASIVE TARGET SCENARIOS IT DOES NOT MATTER WHICH TARGET IS ENGAGED.

DA FORM 7334-R, AUG 2003 APD V1.00

Figure 4-5. Example of completed DA Form 7334-R, TOW Gunnery Tables 3 and 4: Advanced Gunnery Practice and Qualification.

DISK NO.	MISSION NO.
DFT 10	1 Evasive Target/Optical Sight 7 Target Engageability/Optical Sight 12 Fire Control Methods/Thermal Sight 14 Evasive/Obscure Target/Thermal Sight 22 Target Priority/Thermal Sight 24 Target Engageability/Optical 28 Evasive Target/Optical 29 Target Priority/Optical 26 Target Priority/Optical 13 Fratricide/Optical
DFT 12	7 Fratricide/Thermal 9 Fratricide/Thermal 22 Target Priority/Optical 23 Target Priority/Thermal 5 Target Priority/Thermal 6 Target Priority/Thermal 10 Target Priority/Thermal 12 Target Priority/Thermal 15 Evasive Target/Thermal 11 Evasive Target/Thermal
DFT 13	30 Target Engageability/Daysight Tracker 7 Target Priority/Daysight Tracker 26 Target Engageability/Daysight Tracker 21 Fratricide/Daysight Tracker 6 Target Priority/Thermal 10 Target Priority/Thermal 8 Target Priority/Thermal 15 Target Priority/Thermal 12 Target Priority/Thermal
DFT 15	2 Target Engageability/Dual 5 Target Priority/Dual 6 Target Priority/Dual 7 Target Priority/Dual 9 Target Priority/Dual 11 Target Engageability/Dual 12 Evasive Target/Dual 8 15 Target Engageability/Dual 3 15 Target Engageability/Dual 15 Target Engageability/Dual
Note: As reviewer discs become available and are verified, they will be added to the manual as a change.	

Table 4-4. 10-shot events for advanced gunnery tables.

NOTES: 1. In the mission descriptions in Table 4-4, the term "evasive target" refers to evasive, obscured, or difficult targets and helicopters.

2. The term "target priority" refers to missions in which different types of targets are presented (such as tanks and APCs) or in which some are moving and some are stationary. (Using these differences as criteria, the trainer chooses the high priority target for the gunner.)
3. In the missions where the type of sight is listed as dual, both sights are active and the gunner may choose whichever sight he prefers for the engagement or may change sights during the engagement.
4. In the fire control missions, the trainer gives the gunner a choice of fire control methods (frontal, depth, or crossfire), and the gunner must engage the appropriate target out of the group presented.

(5) ***Sustainment Training.*** Each month every soldier assigned to a TOW-equipped unit is required to perform sustainment training on the TGT. The trainer may choose from all missions available for this training. He must remember that he is preparing his soldiers for the quarterly gunnery tables and should choose scenarios that train his soldiers in the skills and engagements in which they are weakest. Since only assigned gunners are *required* to perform and pass TOW Gunnery Tables 3 and 4, the trainer must be sure they are trained on the difficult advanced scenarios. Other members of his TOW crews should be trained as thoroughly as time permits and their training weaknesses require.

e. **TOW Gunnery Table 4, Advanced Gunnery Qualification.** TOW Gunnery Table 4 is similar to TOW Gunnery Table 3. It consists of a series of missions performed with the TGT that train and test advanced gunnery skills. Gunners will be evaluated on their ability to track and hit targets, and on their ability to perform advanced gunnery skills. These skills include determining a target's engageability, identifying targets (to include friend or foe), prioritizing target's, choosing a method of fire control, and engaging targets that are difficult to track, such as helicopters, obscured targets, and extreme-range targets. Only assigned gunners in each squad are required to qualify on TOW Gunnery Table 4. (Other members of the squad may perform the table if time permits and the trainer thinks it useful.) TOW Gunnery Table 4 can only be performed with the TGT. Therefore, units that do not have access to the TGT are not required to perform this table. Gunners must qualify on TOW Gunnery Table 4 to progress to TOW Gunnery Table 5, Baseline TFTT Gunnery Practice.

(1) ***Gunners' Qualification on TOW Gunnery Table 4, Advanced Gunnery Qualification.***

TASK: Qualify on TOW Gunnery Table 4, Advanced Gunnery Qualification.
CONDITIONS: Indoors, using the TGT. The gunner will fire one 10-shot event using the TGT. The 10-shot event will be selected from Table 4-4.
STANDARDS: To qualify on TOW Gunnery Table 4 and progress to TOW Gunnery Table 5, the gunner must score a total of at least 625 points. The precise ranking of the gunner (Expert, 1st class, or 2d class) is determined according to the scale shown on the scorecard. In addition, the gunner must receive a GO on at least 6 out of 10 missions.

---

(2) ***Personnel Duties.*** Personnel duties, conduct of training, and sustainment training are the same as for TOW Gunnery Table 3.

(3) ***Scoring.*** TOW Gunnery Table 4 is scored the same as TOW Gunnery Table 3. The form shown in Figure 4-5, TOW Gunnery Tables 3 and 4, Advanced Gunnery Practice and Qualification Scorecard, is used to score the table.

4-8. PHASE II: SQUAD TRAINING

Phase II begins squad-level TOW training. TOW Gunnery Tables 5 and 6 are static gunnery exercises from a baseline. TOW Gunnery Tables 7 and 8 are moving gunnery exercises where the vehicle moves from point to point on a course and engages a series of targets. These gunnery exercises are used to train and evaluate the TOW unit's combat capability. Each element has two tables: a practice table to train and a qualification table for evaluation. Each table is performed once during the day and once during the night under realistic (simulated) battlefield conditions. Point scores for target engagements are based on a combat model that predicts success based on the time from target detection to target engagement, type of target, and range to target. These scores are taken from the live-fire point calculation sheet (Figure 4-16, 4-35). Point scores are also affected by crew cuts (crew cuts are crew errors, either major or minor, that reduce the point score for a target engagement). The actual rating is calculated by adding the scores for the day and night phases and dividing them by two to get an average score. During move-out scenarios, the senior trainer is normally to the rear of the leader's vehicle (75 meters by 90 degrees) for evaluation and training purposes. An assistant evaluator of the appropriate rank and experience is normally assigned to each additional vehicle that is evaluated as part of the same exercise. Evaluators must monitor crew conversations and fire commands.

a. **TOW Gunnery Table 5, Baseline TFTT Gunnery.** TOW Gunnery Table 5 reacquaints crews with TFTT and allows them to troubleshoot the equipment. It refreshes crews on TFTT gunnery and target engagement techniques to prepare the crews for the practice and qualification tables that follow.

TASK: Conduct baseline TFTT gunnery practice (day and night).

CONDITIONS: On an MPRC or in the field, given a TOW weapon system, its organic carrier, TFTT system, M80 blast simulators, and either pop-up moving targetry or tactical vehicles equipped with TFTT retroreflector assembly, as appropriate. (Use of MILES harness and LTIDs is optional.) The table will be conducted once during daylight and once at night.

STANDARDS: To complete TOW Gunnery Table 5 and progress to TOW Gunnery Table 6, the crew must successfully install, troubleshoot, and operate the TFTT system once during daylight and again at night IAW the appropriate TM. The crew must also successfully acquire, engage, and destroy 6 out of 10 targets presented in Table 5 (once during daylight and once at night).

---

(1) ***Conduct.*** In TOW Gunnery Table 5, the system and or carriers are set up in static positions on a baseline. A range with pop-up or moving targets equipped with TFTT retroreflectors should be used if available. (Use of MILES harness or LTIDs is optional.) If such a range is not available, tactical vehicles equipped with TFTT retroreflectors may be used. (Use of tactical vehicles equipped with MILES harness or LTIDs is optional.) The TOW systems will not move but will engage a series of moving and or pop-up targets from the baseline. TOW Gunnery Table 5 will be fired twice; once during daylight and once at night. Some of the engagements will be designated as NBC engagements and will be performed with the crew in MOPP 4.

(2) ***Scoring.*** To count as a successful engagement, the target must be killed within 30 seconds or less after exposure. Use DA Form 7335-R (TOW Gunnery Tables 5 and 6, Baseline TFTT Gunnery Practice and Qualification Scorecard, Figure 4-6) to record the crew's score (day or night).

(3) ***Suggested Support Requirements.*** The following support requirements are recommended when conducting TOW Gunnery Tables 5 through 8.

- Minimum evaluators (evaluators will not assist the gunner in finding or identifying targets):
  - — One evaluator per squad.
  - — One evaluator in the range tower to record scores on engagements.
- Opposing forces: None.
- Support troops: None.
- Vehicles/communications:
  - — One AN/PRC-119 radio per squad evaluator.
  - — Two AN/PRC-119 radios for the tower evaluator (one primary, one spare).
  - — Two frequencies (one for the tower and squad evaluator and one for the antiarmor squad).
- Maneuver area:
  - — Multipurpose range complex.
  - — A route for the element to move from the assembly area to the first firing point without damaging the underground wiring.
- Firing area:
  - — The TOW backblast area (75 meters by 90 degrees) will be clear.
  - — Smoke may be used.
- Training aids, devices, and special equipment:
  - — TFTT system.
  - — Sufficient TFTT retroreflectors to cover targets. (Use of MILES harness and LTIDs is optional.)
- Ammunition: Ten M80 blast simulators per TOW, per iteration.
- Key references: FM 3-21.91(FM 7-91), FM 3-22.34(FM 23-34), and live-fire point calculation sheet.

(4) ***Trainer/Evaluator Guidelines.*** Because this table is conducted on an MPRC, marking the firing positions may be necessary. At night, the roads must be marked with luminous tape.

b. **TOW Gunnery Table 6, Baseline TFTT Gunnery Qualification.** TOW Gunnery Table 6 is conducted according to the same conditions and standards as TOW

Gunnery Table 5. DA Form 7335-R (TOW Gunnery Tables 5 and 6, Baseline TFTT Gunnery Practice and Qualification Scorecard, Figure 4-6) is used to score the table.

TOW GUNNERY TABLES 5 AND 6: BASELINE TFTT GUNNERY

☐ PRACTICE ☒ QUALIFICATION

For use of this form, see FM 3-22.34. The proponent agency is TRADOC.

SQUAD	SECTION	PLATOON	COMPANY
1ST	1ST	1ST	C 3/60 IN

DATE	DAY	NIGHT
3 SEP 03	X	

ENGAGEMENT	TARGET DESCRIPTION	HIT/MISS/TIME	GO/NG GO
TARGET 1	STATIONARY T72/FLANK/ 1,000 - 1,500 METERS	(HIT) MISS TIME: 21 SEC	GO
TARGET 2	STATIONARY T72/FLANK/ 2,100 - 3,750 METERS	(HIT) MISS TIME: 25 SEC	GO
TARGET 3	STATIONARY T72/FRONTAL/ 2,100 - 3,750 METERS	(HIT) MISS TIME: 23 SEC	GO
TARGET 4	STATIONARY T72/OBLIQUE/ 2,100 - 3,750 METERS	(HIT) MISS TIME: 24 SEC	GO
TARGET 5	STATIONARY T72/DEFILADE/ 800 - 1,500 METERS/NBC	HIT (MISS) TIME: 21 SEC	NG
TARGET 6	MOVING T72/FLANK/ L-R, 30 KPH 1,600 - 2,000 METERS/NBC	(HIT) MISS TIME: 28 SEC	GO
TARGET 7	MOVING T72/FLANK/ L-R, 30 KPH 1,600 - 2,000 METERS/NBC	(HIT) MISS TIME: 26 SEC	GO
TARGET 8	MOVING T72/FLANK/ L-R, 5 KPH 1,000 - 1,500 METERS	(HIT) MISS TIME: 27 SEC	GO
TARGET 9	MOVING T72/FLANK/ L-R, 10 KPH 2,100 - 3,750 METERS	(HIT) MISS TIME: 26 SEC	GO
TARGET 10	MOVING T72/FLANK/ L-R, 40 KPH 1,000 - 1,500 METERS	(HIT) MISS TIME: 23 SEC	GO
		OVERALL GO/NO GO:	9

TO RECEIVE A "GO" ON ANY TARGET ENGAGEMENT, THE GUNNER MUST HIT THE TARGET WITHIN 30 SECONDS AFTER THE TARGET IS EXPOSED. TO RECEIVE A "GO" ON TABLE 5, THE CREW MUST RECEIVE A "GO" ON "INSTALL TFTT" AND A "GO" ON AT LEAST 6 OUT OF 10 TARGET ENGAGEMENTS, BOTH DAY AND NIGHT.

DA FORM 7335-R, AUG 2003 APD V1.00

Figure 4-6. Example of completed DA Form 7335-R, TOW Gunnery Tables 5 and 6: Baseline TFTT Gunnery Practice and Qualification.

c. **TOW Gunnery Table 7, Squad Gunnery Practice.** This table trains the TOW squad to engage stationary and moving targets in a simulated battlefield scenario (Figure 4-7, page 4-22). It prepares the TOW squad for TOW Gunnery Table 8, Squad Gunnery Qualification, by presenting engagement scenarios and task requirements similar to those for evaluation. TOW Gunnery Table 7 may be conducted by a TOW HMMWV on an MPRC light.

(1) ***Conduct.*** All firings will be from the hull down position. Reload procedures will be executed from a defilade position. All stationary targets should be mounted on pop-up mechanisms to facilitate target acquisition and scoring. If pop-up mechanisms are not available, exposed stationary targets will be placed in an appropriate array. If exposed targets must be used, the exercise should be designed so that targets are presented only to the position from which they should be engaged. Moving targets should be presented at speeds from 5 to 40 kilometers per hour.

(2) ***Scoring.*** The squad is rated based on the total point score. For scoring gunnery, time will start for the first target when the target is raised and will stop when the target is hit. All reloads will be made from a hull defilade position. Time for the second target begins when the vehicle reoccupies the firing position and ends when the target is hit. DA Form 7336-R (TOW Gunnery Table 7, Squad Gunnery Practice Scoresheet, Figure 4-8, page 4-23) will be used to record scores.

(a) The table consists of five tasks. Each task consists of one or more target engagements from one firing position. Each target engagement is worth a maximum of 100 points. The point score is determined by the time taken to kill the target using the live-fire point calculation sheet (Figure 4-16, page 4-35). Any crew or leader cuts are deducted from this score (Table 4-5, page 4-21 and Table 4-6, page 4-26).

(b) Each task is also worth a maximum of 100 possible points. This score is calculated by adding the scores for all engagements made during the task and dividing by the number of targets engaged. For example, three targets were engaged and the scores were 75, 82, and 93 for a total of 250 points divided by 3 (the number of targets presented) for a task score of 81 points.

(c) Each table is performed once during the day and once at night. Scoring for day engagements will be made by the evaluator in the range tower. Scoring for night engagements will be made by both evaluators. The scores of the two events are added together for a maximum possible score of 1,000 points. This is the final score for the gunnery table.

NOTE: This scoring process is the same for Tables 7 through 12.

CREW CUTS	
MINOR (Deduct 5 points)	**MAJOR (Deduct 30 points)**
Improper fire commands. Firing before receiving command to fire. Incorrect engagement techniques (such as engaging the least dangerous target before the most dangerous target).	Failure to conduct system self-test/boresight before engaging first target. Failure to collimate day sight and or nightsight before engaging first target. Failure to be in MOPP4 and close all hatches during NBC engagement. Engagement of friendly targets (if friendly targets are presented).

Table 4-5. Crew cuts.

d. **TOW Gunnery Table 8, Squad Gunnery Qualification**. TOW Gunnery Table 8 evaluates the TOW squad on its ability to engage stationary and moving targets in a simulated battlefield scenario (Figure 4-7, page 4-22). The conduct, scoring, suggested support requirements, and trainer/evaluator guidelines are the same as for TOW Gunnery Table 7. The squad's scores will be recorded on DA Form 7337-R, (TOW Gunnery Table 8, Squad Qualification Scoresheet, Figure 4-9, page 4-24).

OPERATION ORDER

SITUATION:

Enemy. The enemy is trying to gain control of key military bases, landing strips, and ports. The 125th and 45th Motorized Rifle Regiments have pushed north towards the objectives of the airfields at Ft. Campbell and Duck AFB. They have been held along the 00 East-West gridline in the vicinity of McKenna. The 125th MPR has swung northeast and is advancing on objectives at EH 42250770 (Watts airstrip) and EH 45101235 (Duffell railhead). Their intent is to secure these areas to facilitae resupply for themselves and for future reinforcements. The 125th is operating at 75 per cent strength and morale is high with the anticipation that they will meet their objectives with little resistance from American forces in the area. The 125th has T72s and BRMs. They are well trained in chemical warfare and have used nonpersistant chemical agents in recent engagements.

Friendly. The section mission is to deny enemy reconnaissances the ability to gain information on the defense of Watts airstrip and Duffell railhead. To our left is __________ (whatever section is appropriate). A Company of the 15th Engineer Battalion has constructed some firing positions and barriers. Our platoon will have priority of fires from one section of our battalion 120mm mortars.

MISSION:

The mission of your squad will be to defend in sector north of Marshall Creek to destroy enemy reconnaissance to prevent them from gaining intelligence concerning Watts airstrip.

EXECUTION:

Concept of the operation. Your squad will occupy on order five battle positions: BP 1 at EH 42260840; BP 2 at EH 42560855; BP 3 at EH 42000858; BP 4 at EH 41810860; and BP 5 at EH 41460872. On order, you will depart from the release point at EH 42340866 and move to BP 1.

Figure 4-7. Example squad gunnery practice and qualification scenario.

TOW GUNNERY TABLE 7: SQUAD GUNNERY PRACTICE							
For use of this form, see FM 3-22.34. The proponent agency is TRADOC.							
TC: SSG MARK LDR: ______ GNR: SPC ERICKSON DVR: SPC MAXWELL				DATE: 5 AUG 03	TIME: 1645		
PLATOON: 2D COMPANY: B				BATTALION: 2/2 IN. LAB			
TASK	CONDITIONS/ TARGETS/ SITUATION	TOTAL TARGET EXPOSURE TIME	HITS	ENGAGEMENT TIMES	POINTS	CREW/ LDR CUTS	TOTAL POINTS
1. ENGAGE A SINGLE TARGET	FIRED FROM BASELINE. 3,000 - 3,750 METERS 1 STATIONARY TANK	TFTT 25 SECS ALL VEHICLES	(1)	ST TK-17	92		92
2. ENGAGE MULTIPLE TARGETS	FIRED FROM PHASE LINE 2. 2,000 - 2,500 METERS 1 - ST TANK, 1 - MV TANK, 1 - MV BMP	TFTT HMMWV/2:30	(1) (2) (3)	ST TK-18 MV TK-40 MV BMP-35	86 83 24		64
3. ENGAGE MULTIPLE TARGETS	FIRED FROM PHASE LINE 3. 2,500 - 3,000 METERS 1 - MV TANK, 1 - ST BMP	TFTT HMMWV/1:40	(1) (2)	ST TK-0 ST BMP-18	0 86	-5 IMPROPER FIRE COMMAND	38
4. ENGAGE MULTIPLE TARGETS	FIRED FROM PHASE LINE 2. 2,500 - 3,000 METERS 1 ST TANK, 1 - MV TANK (NBC)	TFTT HMMWV/1:40	(1) (2)	ST TK-18 MV TK-34	86 94		90
5. ENGAGE MULTIPLE TARGETS	FIRED FROM BASELINE. 2,500 - 3,000 METERS 1 - MV TANK, 1 - MV BMP (NBC)	TFTT HMMWV/1:40	(1) (2)	MV TK-34 MV BMP-18	94 86		60
					TOTAL SCORE		344

NOTES:

1. ENGAGEMENT TIMES ARE CALCULATED FROM THE TIME THE TARGET IS EXPOSED TO THE TIME THE TARGET IS HIT. USE THE LIVE-FIRE POINT CALCULATION SHEET FOR SECOND AND THIRD ENGAGEMENTS FROM THE SAME POSITION. THE TIME BEGINS WHEN THE VEHICLE REOCCUPIES ITS FIRING POSITION (STOPS MOVING) AND ENDS WHEN THE TARGET IS HIT.
2. EACH TASK IS WORTH 100 POINTS. EACH TARGET IS ALSO WORTH 100 POINTS FOR TASKS WITH MULTIPLE ENGAGEMENTS. TOTAL THE NUMBER OF POINTS AND DIVIDE BY THE NUMBER OF TARGETS PRESENTED.
3. CREWLEADER CUTS: SEE FM 3-22.34, CH 4.
4. SCORING TABLE 7: TABLE 7 IS PERFORMED TWICE, ONCE DURING THE DAY AND ONCE AT NIGHT. EACH EVENT IS WORTH A MAX OF 500 POINTS. ADD THE TWO SCORES TOGETHER FOR A POSSIBLE TOTAL OF 1,000 POINTS. SQUADS MUST ACHIEVE A MINIMUM TOTAL OF 700 POINTS TO VERIFY ON TABLE 7 AND PROGRESS TO TABLE 8.

DA FORM 7336-R, AUG 2003 APD V1.00

Figure 4-8. Example of completed DA Form 7336-R, TOW Gunnery Table 7: Squad Gunnery Practice.

TOW GUNNERY TABLE 8: SQUAD GUNNERY QUALIFICATION

For use of this form, see FM 3-22.34. The proponent agency is TRADOC.

TC: SSG MAXWELL LDR: ______ GNR: SPC COOK DVR: PFC ASHLEY | DATE: 6 AUG 03 TIME: 1945

PLATOON: 1ST COMPANY: D BATTALION: 1/502 IN

TASK	CONDITIONS/ TARGETS/ SITUATION	TOTAL TARGET EXPOSURE TIME	HITS	ENGAGEMENT TIMES	POINTS	CREW/ LDR CUTS	TOTAL POINTS
1. ENGAGE A SINGLE TARGET	FIRED FROM BASELINE. 1,500 - 2,000 METERS 1 STATIONARY TANK	TFTT 25 SECS ALL VEHICLES	1	ST TK-20	74		74
2. ENGAGE MULTIPLE TARGETS	FIRED FROM PHASE LINE 1. 2,000 - 2,500 METERS 1 - MV TANK, 1 - ST BMP	TFTT HMMWV/1:40	1 2	MV TK-34 ST BMP-18	94 86		90
3. ENGAGE MULTIPLE TARGETS	FIRED FROM PHASE LINE 2. 3,000 - 3,750 METERS 2 - MV TANK, 1 - ST BMP	TFTT HMMWV/2:45	1 2 3	MV TK-30 MV TK-25 ST BMP-0	94 98 0	-5 IMPROPER FIRE COMMAND	60
4. ENGAGE MULTIPLE TARGETS	FIRED FROM PHASE LINE 1. 1,500 - 2,000 METERS 1 - ST TANK, 1 - MV TANK (NBC)	TFTT HMMWV/1:40	1 2	ST TK-20 MV BMP-19	74 84		79
5. ENGAGE MULTIPLE TARGETS	FIRED FROM BASELINE. 2,500 - 3,000 METERS 1 - MV TANK, 1 - MV BMP (NBC)	TFTT HMMWV/1:40	1 2	MV TK-34 MV BMP-18	94 86	-5 IMPROPER FIRE COMMAND	85
					TOTAL SCORE		388

NOTES:

1. TABLE 8 IS CONDUCTED AND TIMED IN EXACTLY THE SAME MANNER AS TABLE 7.
2. TABLE 8, UNLIKE TABLE 7, IS FOR QUALIFICATION. THE SQUAD IS RATED ON ITS FINAL SCORE, RATHER THAN SIMPLY HAVING TO MEET A MINIMUM STANDARD.
3. CREWLEADER CUTS: SEE FM 3-22.34, CH 4.
4. SCORING TABLE 8: TABLE 8 IS PERFORMED TWICE, ONCE DURING THE DAY AND ONCE AT NIGHT. EACH EVENT IS WORTH A MAX OF 500 POINTS. ADD THE TWO SCORES TOGETHER FOR A POSSIBLE TOTAL OF 1,000 POINTS. SQUADS ARE RATED ON THE FOLLOWING BASIS:

DISTINGUISHED: 900 - 1,000 POINTS SUPERIOR: 800 - 899 POINTS QUALIFIED: 700 - 799 POINTS UNQUALIFIED: 0 - 699 POINTS

DA FORM 7337-R, AUG 2003 APD V1.00

Figure 4-9. Example of completed DA Form 7337-R, TOW Gunnery Table 8: Squad Gunnery Qualification.

4-9. PHASE III: SECTION TRAINING

Phase III begins section-level TOW training. TOW Gunnery Tables 9 and 10 are moving gunnery exercises where the vehicle moves from point to point on a course and engages a series of targets. These gunnery exercises are used to train and evaluate the TOW unit's combat capability. Each element has two tables: a practice table to train and a qualification table for evaluation. Each table is performed once during the day and once during the night under realistic (simulated) battlefield conditions. Point scores for target engagements are based on a combat model that predicts success based on the time from target detection to target engagement, type of target, and range to target. These scores are taken from the live-fire point calculation sheet (Figure 4-16, page 4-35). Point scores are also affected by crew cuts (crew cuts are crew errors, either major or minor, that reduce the point score for a target engagement). The actual rating is calculated by adding the scores for the day and night phases and dividing them by two to get an average score. During move-out scenarios, the senior trainer is normally to the rear of the leader's vehicle for evaluation and training purposes. An assistant evaluator of the appropriate rank and experience is normally assigned to each additional vehicle that is evaluated as part of the same exercise. Evaluators must monitor crew conversations and fire commands.

a. **TOW Gunnery Table 9, Section Gunnery Practice.** This table trains the TOW section to control fire and distribution. It contains offensive and defensive engagements in a simulated battlefield scenario (Figure 4-10, page 4-27). TOW Gunnery Table 9 prepares the TOW section for the qualification table by presenting engagement scenarios and task requirements similar to those for evaluation. It may be conducted by a TOW HMMWV on an MPRC light.

(1) ***Conduct.*** All firings will be from the hull down position. Reload procedures will be executed from a defilade position. All stationary targets should be mounted on pop-up mechanisms to facilitate target acquisition and scoring. If pop-up mechanisms are not available, exposed stationary targets will be placed in an appropriate array. If exposed targets must be used, the exercise should be designed so that targets are presented only to the position from which they should be engaged. Moving targets should be presented at speeds from 5 to 40 kilometers per hour.

(2) ***Scoring.*** The section is rated based on the total point score accumulated. For scoring gunnery, times for engagements will be scored separately for each vehicle. Start time for the section leader's vehicle will be the same as in TOW Gunnery Table 7. Start time for the squad leader's vehicle will be upon completion of the fire directive by the section leader, which will include the method of engagement. Stop time will be when the target is hit. Points will be awarded IAW the live-fire point calculation sheet (Figure 4-16, page 4-35). Deductions will be made from this point score when a major or minor crew cut error (Table 4-5, page 4-21) or a section leader cut error (Table 4-6, page 4-26) is made. The evaluator in the range tower will score day engagements. Both evaluators will score night engagements. DA Form 7338-R (TOW Gunnery Table 9, Section Gunnery Practice Scoresheet, Figure 4-11, page 4-28) will be used to record the scores.

LEADER CUTS	
MINOR (Deduct 10 points)	MAJOR (Deduct 30 points)
Give incorrect fire commands. Fail to give SITREP to tower.	— Fail to properly distribute and control section fires.

Table 4-6. Section leader cuts.

(3) ***Suggested support requirements.*** The following support requirements are recommended when conducting TOW Gunnery Tables 9 and 10.

- Minimum evaluators (evaluators will not assist the gunner in finding or identifying targets):
  — One evaluator per section and squad.
  — Two evaluators in the range tower to record scores on engagements.
- Opposing forces: None.
- Support troops: None.
- Vehicles/communications:
  — One AN/PRC-119 radio per section and squad evaluator.
  — Two AN/PRC-119 radios for the tower evaluator (one primary, one spare).
  — Two frequencies (one for the tower and section evaluator and one for the antiarmor squad).
- Maneuver area:
  — Multipurpose range complex.
  — A route for the element to move from the assembly area to the first firing point without damaging the underground wiring.
- Firing area:
  — The TOW backblast area (75 meters by 90 degrees) will be clear.
  — Smoke may be used.
- Training aids, devices, and special equipment:
  — TFTT system.
  — Sufficient TFTT retroreflectors to cover the targets. (Use of MILES harness and LTIDs is optional.)
- Ammunition: Six M80 blast simulators per TOW, per iteration.
- Key references: FM 3-21.91(FM 7-91), FM 3-22.34(FM 23-34), and live-fire point calculation sheet.

(4) ***Trainer/Evaluator Guidelines***. Because this table is conducted on an MPRC, marking the firing positions may be necessary. At night, the roads must be marked with luminous tape.

b. **TOW Gunnery Table 10, Section Gunnery Qualification.** TOW Gunnery Table 10 evaluates the TOW section on controlling fire and distribution. It contains offensive and defensive engagements in a simulated battlefield scenario (Figure 4-10, page 4-27). The conduct, scoring, suggested support requirements, and trainer/evaluator guidelines are the same as for TOW Gunnery Table 9. The section's scores will be recorded on

DA Form 7339-R (TOW Gunnery Table 10, Section Gunnery Qualification Scoresheet, Figure 4-12, page 4-29).

OPERATION ORDER

SITUATION:

Enemy. The enemy is trying to gain control of key military bases, landing strips, and ports. The 125th and 45th Motorized Rifle Regiments have pushed north towards the objectives of the airfields at Ft. Campbell and Duck AFB. They have been held along the 00 East-West gridline in the vicinity of McKenna. The 125th MRR has swung northeast and is advancing on objectives at EH 42250770 (Watts airstrip) and EH 45101235 (Duffell railhead). Their intent is to secure these areas to facilitate resupply for themselves and for future reinforcements. The 125th is operating at 75 per cent strength and morale is high with the anticipation that they will meet their objectives with little resistance from American forces in the area. The 125th has T72s and BRMs. They are well trained in chemical warfare and have used nonpersistant chemical agents in recent engagements.

Friendly. The platoon mission is to deny enemy reconnaissances the ability to gain information on the defense of Watts airstrip and Duffell railhead. To our left is __________ (whatever platoon is appropriate). A Company of the 15th Engineer Battalion hs constructed some firing positions and barriers. Our platoon will have priority of fires from one section of our battalion 120mm mortars.

MISSION:

The mission of your section will be to defend in sector north of Marshall Creek to destroy enemy reconnaissance to prevent them from gaining intelligence concerning Watts airstrip.

EXECUTION:

Concept of the operation. Your section will occupy on order five battle positions: BP 1 at EH 42260840; BP 2 at EH 42560855; BP 3 at EH 42000858; BP 4 at EH 41810860; and BP 5 at EH 41460872. On order, you will depart from the release point at EH 42340866 and move to BP 1.

Figure 4-10. Example section gunnery practice and qualification scenario.

TOW GUNNERY TABLE 9: SECTION GUNNERY PRACTICE

For use of this form, see FM 3-22.34. The proponent agency is TRADOC.

SECTION	TC: SSG HEIDER	LDR:	GNR: SPC CONNOR	DVR: PFC BROWN	DATE:	15 AUG 03
SQUAD	TC: SGT QUEEN	LDR:	GNR: PFC MARKUS	DVR: PFC WILSON	TIME:	0900

PLATOON: ______ COMPANY: ______ BATTALION: ______

TASK	CONDITIONS/ TARGETS/ SITUATION	TOTAL TARGET EXPOSURE TIME	HITS	ENGAGEMENT TIMES	POINTS	CREW/ LDR CUTS	TOTAL POINTS
1. ENGAGE A SINGLE TARGET	FIRED FROM BASELINE. 3,000 - 3,750 METERS 1 - ST TANK, 1 - ST BMP	TFTT 40 SECS ALL VEHICLES	① ②	ST - 20 SB - 30	74 35		55
2. ENGAGE MULTIPLE TARGETS	FIRED FROM PHASE LINE 2. 2,000 - 2,750 METERS 1 - ST BMP	TFTT 25 SECS ALL VEHICLES	①	SB - 20	74		74
3. ENGAGE MULTIPLE TARGETS	FIRED FROM PHASE LINE 3. 1,000 - 1,500 METERS 2 - MV BMPS, 1 - ST TANK	TFTT HMMWV/1:40	① ② ③	MB - 16 MB - 18 ST - 13	91 87 100		93
4. ENGAGE MULTIPLE TARGETS	FIRED FROM PHASE LINE 2. 2,500 - 3,000 METERS 2 - MV TANKS, 2 - ST TANKS (NBC)	TFTT HMMWV/2:38	① ② ③ ④	MT - 19 MT - 21 ST - 20 ST - 22	100 99 74 64	-10 INCORRECT FIRE COMMAND	74
5. ENGAGE MULTIPLE TARGETS	FIRED FROM BASELINE. 1,000 - 1,500 METERS 1 - ST TANK, 1 - ST BMP (NBC)	TFTT 25 SECS ALL VEHICLES	① ②	ST - 16 SB - 17	85 92		89
					TOTAL SCORE		385

NOTES:

1. TABLE 9 IS CONDUCTED AND TIMED IN EXACTLY THE SAME MANNER AS TABLES 7 AND 8.
2. SCORING TABLE 9: TABLE 9 IS PERFORMED TWICE, ONCE DURING THE DAY AND ONCE AT NIGHT. EACH EVENT IS WORTH A MAX OF 500 POINTS. ADD THE TWO SCORES TOGETHER FOR A POSSIBLE TOTAL OF 1,000 POINTS. TABLE 9 IS A PRACTICE TABLE. TO VERIFY ON TABLE 8 AND PROGRESS TO TABLE 10, THE SECTION MUST ACHIEVE A MINIMUM COMBINED SCORE OF AT LEAST 700 POINTS, AFTER CREW AND LEADER CUTS.

DA FORM 7338-R, AUG 2003 APD V1.00

Figure 4-11. Example of completed DA Form 7338-R, TOW Gunnery Table 9: Section Gunnery Practice.

TOW GUNNERY TABLE 10: SECTION GUNNERY QUALIFICATION

For use of this form, see FM 3-22.34. The proponent agency is TRADOC.

SECTION TC: SGT BERRY LDR: PFC KIM GNR: SPC MARKS DVR: PFC SMITH DATE: 28 AUG 03

SQUAD TC: SGT WATT LDR: PFC LUCAS GNR: SPC COLEMAN DVR: PFC RICHARDS TIME: 1600

PLATOON: 2D COMPANY: E BATTALION: 2/16 IN

TASK	CONDITIONS/ TARGETS/ SITUATION	TOTAL TARGET EXPOSURE TIME	HITS	ENGAGEMENT TIMES	POINTS	CREW/ LDR CUTS	TOTAL POINTS
1. ENGAGE A SINGLE TARGET	FIRED FROM BASELINE. 3,000 - 3,750 METERS 2 - MV TANK, 2 - ST BMPS	TFTT HMMWV/2:38	1 2 3 4	MT-30 MT-37 SB-17 SB-16	96 93 92 100	-10 INCORRECT ENGAGEMENT TECH.	85
2. ENGAGE MULTIPLE TARGETS	FIRED FROM PHASE LINE 2. 2,000 - 2,500 METERS 1- MV TANK	TFTT 40 SECS ALL VEHICLES	1	MT-40	83		83
3. ENGAGE MULTIPLE TARGETS	FIRED FROM PHASE LINE 3. 1,000 - 1,500 METERS 1 - MV BMP, 1 - ST TANK	TFTT 25 SECS ALL VEHICLES	1 2	ST-64 MB-59	64 59		62
4. ENGAGE MULTIPLE TARGETS	FIRED FROM PHASE LINE 2. 2,500 - 3,000 METERS 1 - ST BMP, 2 - MV TANKS (NBC)	TFTT HMMWV/2:40	1 2 3	SB-18 MT-34 MT-24	86 94 98		93
5. ENGAGE MULTIPLE TARGETS	FIRED FROM BASELINE. 3,000 - 3,750 METERS 1 - ST TANK, 1 - ST BMP (NBC)	TFTT 25 SECS ALL VEHICLES	1 2	ST-20 MB-0	74 0		37
					TOTAL SCORE		360

NOTES:

1. TABLE 10 IS CONDUCTED AND TIMED IN EXACTLY THE SAME MANNER AS TABLE 8 AND 9.
2. SCORING TABLE 10: TABLE 10 IS PERFORMED TWICE, ONCE DURING THE DAY AND ONCE AT NIGHT. EACH EVENT IS WORTH A MAX OF 500 POINTS. ADD THE TWO SCORES TOGETHER FOR A POSSIBLE TOTAL OF 1,000 POINTS. TABLE 10 IS A QUALIFICATION TABLE. TO PROGRESS TO TABLE 11, THE SECTION MUST ACHIEVE A MINIMUM COMBINED SCORE OF AT LEAST 700 POINTS.

DA FORM 7339-R, AUG 2003 APD V1.00

Figure 4-12. Example of completed DA Form 7339-R, TOW Gunnery Table 10: TOW Section Gunnery Qualification.

4-10. PHASE IV: PLATOON TRAINING

Phase IV begins platoon-level TOW training. TOW Gunnery Tables 11 and 12 are moving gunnery exercises where the vehicle moves from point to point on a course and engages a series of targets. These gunnery exercises are used to train and evaluate the TOW unit's combat capability. Each element has two tables: a practice table to train and a qualification table for evaluation. Each table is performed once during the day and once during the night under realistic (simulated) battlefield conditions. Point scores for target engagements are based on a combat model that predicts success based on the time from target detection to target engagement, type of target, and range to target. These scores are taken from the live-fire point calculation sheet (Figure 4-16, page 4-35). Point scores are also affected by crew cuts (crew cuts are crew errors, either major or minor, that reduce the point score for a target engagement). The actual rating is calculated by adding the scores for the day and night phases and dividing them by two to get an average score. During move-out scenarios, the senior trainer is normally to the rear of the leader's vehicle for evaluation and training purposes. An assistant evaluator of the appropriate rank and experience is normally assigned to each additional vehicle that is evaluated as part of the same exercise. Evaluators must monitor crew conversations and fire commands.

a. **TOW Gunnery Table 11, Platoon Gunnery Practice.** This table trains the TOW platoon to control fire and distribution. It contains offensive and defensive engagements in a simulated battlefield scenario (Figure 4-13, page 4-32). TOW Gunnery Table 11 prepares the TOW platoon for the qualification table by presenting engagement scenarios and task requirements similar to those for evaluation. It may be conducted by a TOW HMMWV on an MPRC light.

(1) ***Conduct.*** All firings will be from the hull down position. Reload procedures will be executed from a defilade position. All stationary targets should be mounted on pop-up mechanisms to facilitate target acquisition and scoring. If pop-up mechanisms are not available, exposed stationary targets will be placed in an appropriate array. If exposed targets must be used, the exercise should be designed so that targets are presented only to the position from which they should be engaged. Moving targets should be presented at speeds from 5 to 40 kilometers per hour.

(2) ***Scoring.*** The platoon is rated based on the total point score accumulated. For scoring gunnery, times for engagements will be scored separately for each vehicle. Start time will begin upon completion of the directive given to the vehicle commander. Stop time will be when the target is hit. Points will be awarded IAW the live-fire point calculation sheet (Figure 4-16, page 4-35). Deductions will be made from this point score when a major or minor crew cut error (Table 4-5, page 4-21) or a platoon leader cut error (Table 4-7) is made. The evaluator in the range tower will score day engagements. Both evaluators will score night engagements. DA Form 7340R (TOW Gunnery Table 11, Platoon Gunnery Practice Scoresheet, Figure 4-14, page 4-33) will be used to record the scores.

PLATOON LEADER CUTS	
MINOR (Deduct 10 points)	MAJOR (Deduct 30 points)
Give incorrect fire commands. Fail to give SITREP to tower.	Fail to properly distribute and control section fires.

Table 4-7. Platoon leader cuts.

(3) ***Suggested Support Requirements.*** The following support requirements are recommended when conducting TOW Gunnery Tables 11 and 12.

- Minimum evaluators (evaluators will not assist the gunner in finding or identifying targets):
  - — One evaluator per platoon, section, and squad.
  - — Two evaluators in the range tower to record scores on engagements.
- Opposing forces: None.
- Support troops: None.
- Vehicles/communications:
  - — One AN/PRC-119 radio per platoon, section, and squad evaluator.
  - — Two AN/PRC-119 radios for the tower evaluator (one primary, one spare).
  - — Two frequencies (one for the tower and platoon evaluator and one for the antiarmor platoon).
- Maneuver area:
  - — Multipurpose range complex.
  - — A route for the element to move from the assembly area to the first firing point without damaging the underground wiring.
- Firing area:
  - — The TOW backblast area (75 meters by 90 degrees) will be clear.
  - — Smoke may be used.
- Training aids, devices, and special equipment:
  - — TFTT system.
  - — Sufficient TFTT retroreflectors to cover the targets. (Use of MILES harness and LTIDs is optional.)
- Ammunition: Six M80 blast simulators per TOW, per iteration.
- Key references: FM 3-21.91(FM 7-91), FM 3-22.34(FM 23-34), and live-fire point calculation sheet.

(4) ***Trainer/Evaluator Guidelines.*** Because this table is conducted on an MPRC, marking the firing positions may be necessary. At night, the roads must be marked with luminous tape.

b. **TOW Gunnery Table 12, Platoon Gunnery Qualification.** TOW Gunnery Table 12 evaluates the TOW platoon on controlling fire and distribution. It contains offensive and defensive engagements in a simulated battlefield scenario (Figure 4-13, page 4-32). The conduct, scoring, suggested support requirements, and trainer/evaluator guidelines are the same as for TOW Gunnery Table 11. The platoon's scores will be recorded on DA

Form 7341-R (TOW Gunnery Table 12, Platoon Gunnery Qualification Scoresheet, Figure 4-15, page 4-34).

SITUATION:

Enemy. The enemy is trying to gain control of key military bases, landing strips, and ports. The 125th and 45th Motorized Rifle Regiments have pushed north towards the objectives of the airfields at Ft. Campbell and Duck AFB. They have been held along the 00 East-West gridline in the vicinity of McKenna. The 125th MRR has swung northeast and is advancing on objectives at EH 42250770 (Watts airstrip) and EH 45101235 (Duffell railhead). Their intent is to secure these areas to facilitate resupply for themselves and for future reinforcements. The 125th is operating at 75 per cent strength and morale is high with the anticipation that they will meet their objectives with little resistance from American forces in the area. The 125th has T72s and BRMs. They are well trained in chemical warfare and have used nonpersistant chemical agents in recent engagements.

Friendly. The company mission is to deny enemy reconnaissances the ability to gain information on the defense of Watts airstrip and Duffell railhead. To our left is __________ (whatever company is appropriate). A Company of the 15th Engineer Battalion hs constructed some firing positions and barriers. Our platoon will have priority of fires from one section of our battalion 120mm mortars.

MISSION:

The mission of your platoon will be to defend in sector north of Marshall Creek to destroy enemy reconnaissance to prevent them from gaining intelligence concerning Watts airstrip.

EXECUTION:

Concept of the operation. Your platoon will occupy on order five battle positions: BP 1 at EH 42260840; BP 2 at EH 42560855; BP 3 at EH 42000858; BP 4 at EH 41810860; and BP 5 at EH 41460872. On order, you will depart from the release point at EH 42340866 and move to BP 1.

Figure 4-13. Example platoon gunnery practice and qualification scenario.

TOW GUNNERY TABLE 11: PLATOON GUNNERY PRACTICE

For use of this form, see FM 3-22.34. The proponent agency is TRADOC.

SECTION 1:	TOW CREW 1:	TC: SSG CAMPBELL	LDR:	GNR: SPC COOK	DVR: PFC BROWN	DATE: 15 AUG 93
	TOW CREW 2:	TC: SGT HAMMON	LDR:	GNR: SPC ADAMS	DVR: PFC GONZALES	
SECTION 2:	TOW CREW 1:	TC: SSG QUEEN	LDR:	GNR: SPC McCOY	DVR: PFC ZALAS	
	TOW CREW 2:	TC: SGT WEBB	LDR:	GNR: SPC COOKS	DVR: PFC SMITH	TIME: 1100
SECTION 3:	TOW CREW 1:	TC: SSG PAINE	LDR:	GNR: SPC ERVIC	DVR: PFC FOWLER	
	TOW CREW 2:	TC: SGT MILLER	LDR:	GNR: SPC SHUTT	DVR: PFC ANDERSON	

PLATOON: 3RD COMPANY: D BATTALION: 2/503 IN

TASK	CONDITIONS/ TARGETS/ SITUATION	TOTAL TARGET EXPOSURE TIME	HITS	ENGAGEMENT TIMES	POINTS	CREW/ LDR CUTS	TOTAL POINTS
1. ENGAGE A SINGLE TARGET	FIRED FROM BASELINE. 3,000 - 3,750 METERS 4 - MV TANKS, 2 - ST BMPS	TFTT HMMWV/3:26	1 2 3 4 5 6	MT - 33 MT - 29 MT - 37 MT - 31 SB - 22 SB - 17	95 97 93 96 64 92		90
2. ENGAGE MULTIPLE TARGETS	FIRED FROM PHASE LINE 2. 2,000 - 3,000 METERS 4 - MV BMPS 2 - ST BMPS	TFTT HMMWV/2:26	1 2 3 4 5 6	MB - 20 MB - 19 MB - 23 MB - 25 SB - 17 SB - 0	73 93 86 82 92 0		71
3. ENGAGE MULTIPLE TARGETS	FIRED FROM PHASE LINE 3. 1,000 - 2,000 METERS 2 - ST TANKS, 2 - ST BMPS	TFTT 25 SECS ALL VEHICLES	1 2 3 4	ST - 17 ST - 19 SB - 21 SB - 26	92 79 68 47		72
4. ENGAGE MULTIPLE TARGETS	FIRED FROM PHASE LINE 2. 2,000 - 3,000 METERS 2- MV TANKS, 2 - MV BMPS (NBC)	TFTT BMP/43 SEC TANK/1:03	1 2 3 4	MT - 17 MT - 22 MB - 23 MB - 19	100 99 86 93	−10 INCORRECT FIRE COMMAND	85
5. ENGAGE MULTIPLE TARGETS	FIRED FROM BASELINE. 3,000 - 3,750 METERS 2 - ST TANKS, 2 - ST BMPS (NBC)	TFTT 25 SECS ALL VEHICLES	1 2 3 4	ST - 15 ST - 21 SB - 24 SB - 0	100 68 55 0		56
						TOTAL SCORE	374

NOTES:

1. TABLE 11 IS CONDUCTED, TIMED, AND SCORED IN EXACTLY THE SAME MANNER AS TABLES 7 THRU 10.

DA FORM 7340-R, AUG 2003

APD V1.00

Figure 4-14. Example of completed DA Form 7340-R, TOW Gunnery Table 11: Platoon Gunnery Practice.

TOW GUNNERY TABLE 12: PLATOON GUNNERY QUALIFICATION

For use of this form, see FM 3-22.34. The proponent agency is TRADOC.

SECTION 1:	TOW CREW 1:	TC: SSG COOKS	LDR: SPC SMITH	GNR: SPC BROOKS	DVR: PFC MIDDLEBROOK	DATE: 15AUG03
	TOW CREW 2:	TC: CPL YORK	LDR: SPC WALTER	GNR: SPC CARTER	DVR: PFC PORTER	
SECTION 2:	TOW CREW 1:	TC: SSG ASHBY	LDR: SPC HART	GNR: SPC JORDAN	DVR: PFC JOHNSON	
	TOW CREW 2:	TC: SGT OH	LDR: PFC CLAY	GNR: SPC PRICE	DVR: PFC CARDEN	TIME: 1400
SECTION 3:	TOW CREW 1:	TC: SGT BLAND	LDR: SPC OLIVER	GNR: SPC SELL	DVR: PFC OGLE	
	TOW CREW 2:	TC: SGT WALKER	LDR: PFC HARRIS	GNR: SPC NOEL	DVR: SPC PETAIA	

PLATOON: 2D COMPANY: E BATTALION: 5/16 IN

TASK	CONDITIONS/ TARGETS/ SITUATION	TOTAL TARGET EXPOSURE TIME	HITS	ENGAGEMENT TIMES	POINTS	CREW/ LDR CUTS	TOTAL POINTS
1. ENGAGE A SINGLE TARGET	FIRED FROM BASELINE. 3,000 - 3,750 METERS 2 - MV TANKS, 2 - ST BMPS	TFTT BMP/25 SECS TANK/1:13	(1) 2 (3) (4)	MT-38 MT-0 SB-19 SB-23	92 0 79 59		58
2. ENGAGE MULTIPLE TARGETS	FIRED FROM PHASE LINE 2. 2,000 - 3,000 METERS 3 - ST BMPS, 3 - MV TANKS	TFTT HMMWV/2:28	(1) (2) (3) (4) (5) (6)	SB-20 SB-17 SB-24 MT-23 MT-33 MT-0	74 92 64 96 90 0	INCORRECT FIRE COMMAND -10	59
3. ENGAGE MULTIPLE TARGETS	FIRED FROM PHASE LINE 3. 1,000 - 2,000 METERS 3 - ST TANKS, 3 - MV BMPS	TFTT HMMWV/1:08	(1) (2) (3) (4) (5) 6	ST-17 ST-19 ST-26 MB-31 MB-28 MB-0	42 79 47 57 61 0		56
4. ENGAGE MULTIPLE TARGETS	FIRED FROM PHASE LINE 2. 1,500 - 2,500 METERS 2- MV TANKS, 2 - MV BMPS (NBC)	TFTT BMP/34 SECS TANK/1:03	(1) (2) (3) 4	MT-23 MT-26 MB-15 MB-0	96 95 100 0		73
5. ENGAGE MULTIPLE TARGETS	FIRED FROM BASELINE. 3,000 - 3,750 METERS 2 - ST TANKS, 2 - ST BMPS (NBC)	TFTT 25 SECS ALL VEHICLES	(1) (2) (3) (4)	ST-15 ST-17 SB-17 SB-23	100 92 42 59	-10 NO SITREP SENT	76
					TOTAL SCORE		322

NOTES:

1. TABLE 12 IS CONDUCTED, TIMED, AND SCORED IN EXACTLY THE SAME MANNER AS TABLES 7 THRU 10.

DA FORM 7341-R, AUG 2003 APD V1.00

Figure 4-15. Example of completed DA Form 7341-R, TOW Gunnery Table 12: TOW Platoon Gunnery Qualification.

		MOVING TARGET					
TIME SEC	STATIONARY TARGETS	BMP 65-900 METERS	BMP 901-2000 METERS	BMP 2001-3500 METERS	TANK 65-1500 METERS	TANK 1501-2500 METERS	TANK 2501-3750 METERS
1	100	100	100	100	100	100	100
2	100	100	100	100	100	100	100
3	100	100	100	100	100	100	100
4	100	100	100	100	100	100	100
5	100	100	100	100	100	100	100
6	100	100	100	100	100	100	100
7	100	100	100	100	100	100	100
8	100	100	100	100	100	100	100
9	100	100	100	100	100	100	100
10	100	100	100	100	100	100	100
11	100	92	100	100	100	100	100
12	100	86	100	100	97	100	100
13	100	79	97	100	94	99	100
14	100	74	95	100	91	99	100
15	100	68	93	100	88	99	100
16	100	64	91	98	85	98	100
17	92	59	89	96	81	98	100
18	86	55	87	95	78	98	100
19	79	51	84	93	75	97	100
20	74	47	82	91	72	97	100
21	68	44	80	89	69	97	99
22	64	41	77	88	66	96	99
23	59	38	75	86	63	96	99
24	55	35	73	84	60	95	98
25	51	33	70	82	57	95	98
26	47	30	68	80	54	94	98
27	44	28	66	79	52	94	97
28	41	26	63	77	49	93	97
29	38	24	61	75	46	92	97
30	35	22	59	73	44	92	96
31	33	21	57	71	41	91	96
32	30	19	54	70	39	90	95
33	28	18	52	68	37	90	85
34	26	16	50	66	34	89	94
35	24	15	48	64	32	88	94
36	22	14	46	62	30	87	93
37	21	13	44	61	28	86	93
38	19	12	42	59	27	85	92
39	18	11	40	57	25	84	92
40	16	10	38	56	23	83	91
41	15	9	36	54	22	82	91
42	14	9	35	52	20	81	90
43	13	8	33	51	19	80	90
44	12	7	31	49	18	79	89
45	11	7	30	47	17	78	89
46	10	6	28	46	15	77	88
47	9	6	27	44	14	75	88
48	9	5	26	43	13	74	87
49	8	5	24	41	12	73	86
50	7	4	23	40	12	71	85
51	7	4	22	39	11	70	84
52	6	4	21	37	11	68	83
53	6	3	19	36	9	67	82
54	5	3	18	35	9	65	81

Figure 4-16. Example of TOW live-fire point calculation sheet.

MOVING TARGET							
TIME SEC	STATIONARY TARGETS	BMP 65-900 METERS	BMP 901-2000 METERS	BMP 2001-3500 METERS	TANK 65-1500 METERS	TANK 1501-2500 METERS	TANK 2501-3750 METERS
55	5	3	17	33	8	64	80
56	4	3	16	32	7	62	78
57	4	2	16	31	7	61	77
58	4	2	15	30	6	59	75
59	3	2	14	29	6	58	74
60	3	2	13	28	5	56	72
61	3	2	12	27	5	54	70
62	3	1	12	25	5	53	69
63	2	1	11	24	4	51	67
64	2	1	10	24	4	49	65
65	2	1	10	23	4	48	63
66	2	1	9	22	3	46	61
67	2	1	9	21	3	45	60
68	1	1	8	20	3	43	58
69	1	1	8	19	2	41	56
70	1	1	7	18	2	40	55
71	1		7	18	2	38	53
72	1		6	17	2	37	52
73	1		6	16	2	35	50
74	1		5	15	2	34	49
75	1		5	15	1	32	47
76	1		5	14	1	31	46
77			5	13	1	30	45
78			4	13	1	28	43
79			4	12	1	27	42
80			4	12	1	26	41
81			3	11	1	25	40
82			3	11	1	23	39
83			3	10	1	22	38
84			3	10	1	21	37
85			3	9		20	36
86			2	9		19	35
87			2	9		18	34
88			2	8		17	33
89			2	8		16	32
90			2	7		15	31
91			2	7		15	30
92			2	7		14	29
93			1	6		13	28
94			1	6		12	27
95			1	5		12	26
96			1	5		11	25
97			1	5		10	24
98			1	5		10	23
99			1	5		9	22
100			1	5		9	21
101			1	4		8	20
102			1	4		8	19
103			1	4		7	18
104				4		7	17
105				3		6	16
106				3		6	15
107				3		6	14
108				3		5	13

Figure 4-16. Example of TOW live-fire point calculation sheet (continued).

4-11. FIELD TRACKING

Field tracking (or tracking in a field environment) provides practice and experience in tracking uncooperative targets and should be performed in conjunction with FTXs. It is not an element in the 12 required TOW gunnery tables, but it is useful in preparing for them, especially Tables 7 through 12. Other tactical employment tasks, such as occupying a firing position or completing a range card, should be performed at the same time. Careful planning by unit commanders will result in effective training that makes the best use of time, equipment, and personnel.

a. The tracking range should include an area more than 3,000 meters deep and at least 500 meters wide with hills and valleys, dead space, and covered terrain. Battlefield target conditions should be simulated as follows:

(1) Targets should be a variety of tanks, APCs, and other tactical vehicles. Many training installations now have various types of actual threat vehicles. These vehicles should be used when possible.

(2) Target vehicles should behave as much like real enemy vehicles on the battlefield as possible. They should change directions constantly and quickly (to include backing up for short distances) and should vary their speed. They should move within the gunner's sight picture from left to right, top to bottom, and bottom to top. The vehicles should move at angles to and from the weapon position as well as directly toward and away from the weapon.

(3) Gunners should see tanks and APCs in full and partial frontal, flank, and rear exposures.

(4) Gunners must engage the targets at all ranges with both the daysight tracker and the nightsight. To do so, they must be able to determine if a target is in range. Targets should move in and out of range so gunners can gain experience in target range determination.

(5) Armored vehicles should limit their exposure by moving quickly from one covered area to another to escape ATGM fires. The squad leader must have the gunner acquire and fire at the target while it is exposed. During field tracking, target vehicle exposure time is reduced to a minimum by having targets use covered areas during halts. This gives the gunner experience in tracking targets that disappear and reappear.

b. Field tracking should be performed under conditions that normally occur on the battlefield. These conditions include limited visibility caused by darkness, smoke, or fog; the discomfort of MOPP gear; and the noise and distraction caused by indirect and small-arms fire around the firing positions. Battlefield tracking conditions should be simulated as follows:

(1) The TOW crew members should train to engage targets during reduced visibility conditions using both the daysight tracker and the nightsight. Training should be conducted during daylight using the AN/TAS-4A. The AN/TAS-4A is a thermal sight, ***not*** just a nightsight. It offers many advantages in searching for and tracking targets in all kinds of weather and at any time of day. The AN/TAS-4A should be used extensively in conjunction with the daysight to maximize the abilities of the system.

(2) Enemy capabilities include the use of nuclear, biological, and chemical munitions. These munitions may not affect the weapon, but a gunner's ability to track will be degraded if he has never tracked while wearing a protective mask and protective clothing. Training is more realistic when the entire crew wears protective masks and clothing during part of the field tracking exercises.

(3) The enemy will try to suppress TOW fires with artillery and mortar fires. A gunner's ability to keep the sight crosshairs on a target may be affected under these conditions, especially if he has not been properly trained. Distractors should be used in

training to accustom gunners to such conditions and to minimize flinching. Grenade and artillery simulators are effective distractors. Small-arms fire can also be expected, but gunners should not be seriously distracted by its noise. Therefore, the simulation of small-arms fire is not required.

c. If facilities and equipment are not available to conduct a field tracking exercise, gunners should be allowed to track any uncooperative targets such as military or civilian vehicles traveling on a nearby highway. Any means available should be used to allow gunners to practice tracking targets that vary target speeds and directions of movement. This should be done often and can be accomplished in any location that has vehicle traffic.

d. TOW crews must be proficient in combat vehicle identification to gain the most benefit from field tracking. Combat vehicle identification skills have been standardized by GTA 17-02-011. This GTA includes six modules of 35-mm slides and instructor guides. Each module highlights five different NATO or Threat combat vehicles from five different views: frontal, right side, left side, right oblique, and left oblique.

CHAPTER 5

COMBAT TECHNIQUES OF FIRE

This chapter discusses the techniques and procedures of fire control, helicopter engagement, and firing the TOW under NBC and limited visibility conditions in a combat situation. These techniques and procedures greatly enhance the performance of the TOW weapon system in combat and increase its chances of survival.

Section I. FIRE CONTROL MEASURES

This section discusses fire control measures in combat to include target engagement determination procedures, the elements of fire commands, target tracking procedures, and target engagement procedures with the M220A1 and M220A2 TOW launchers.

5-1. TARGET ENGAGEMENT DETERMINATION

Mechanized units are trained to use terrain driving techniques to conceal movement, and drivers are taught to move vehicles quickly from one concealed position to another. TOW gunners and squad leaders must determine the range to a target and determine if the exposure time—the time a vehicle is in an opening between positions—is long enough to allow a missile to reach its target.

a. **Determine if a Target is Within Range**. The TOW gunner or squad leader can use the nightsight method or the binocular method to determine if a target is within range.

(1) ***Nightsight Method.*** The reticle within the nightsight is marked with stadia lines on each side of the vertical crossline and just above the intersection of the vertical and horizontal crosslines (Figure 5-1, page 5-2). Use these stadia lines to determine if a target is within range.

(a) Check to ensure the nightsight is in the narrow field of view.

(b) Look at the flank (broadside) of a target. If the ends of the target touch or extend beyond the stadia lines, the target is within range (Figure 5-2, page 5-2).

(c) If the target is moving straight or at an angle toward the position, use one-half of the stadia lines. If one side of the vehicle touches one of the stadia lines and the other side of the vehicle touches or extends beyond the vertical line of the crossline, the target is within range (Figure 5-3, page 5-2).

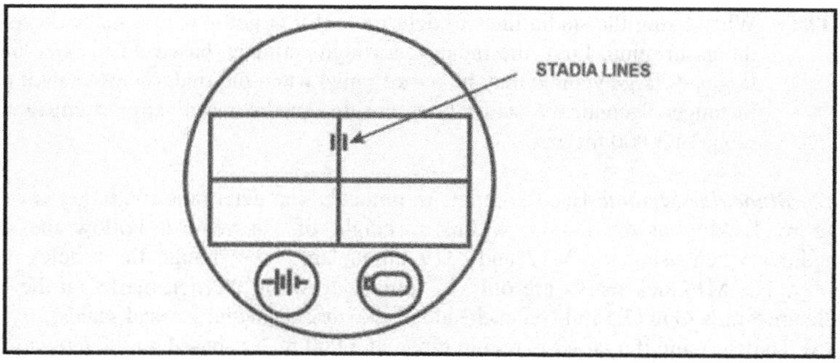

Figure 5-1. Stadia lines.

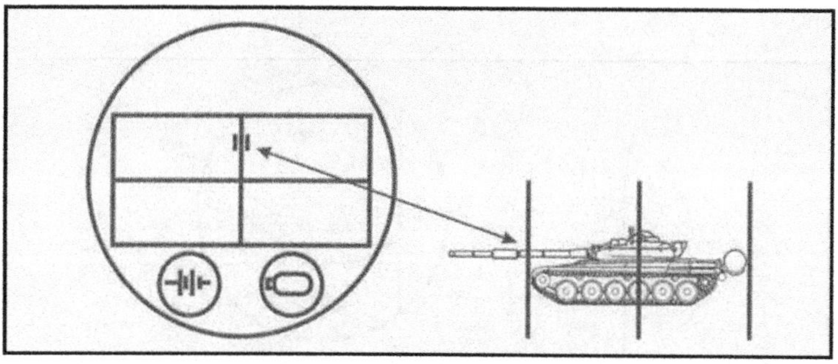

Figure 5-2. Target within range.

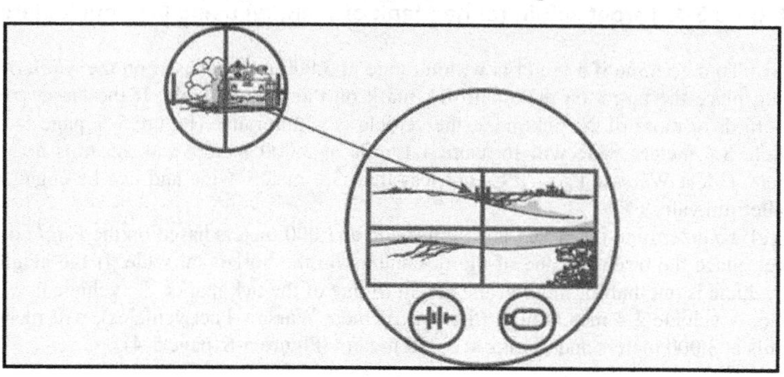

Figure 5-3. Target moving straight toward or away from position within range.

NOTE: When using the stadia lines to determine if a target is within range, keep two things in mind. First, the range is a rough estimate, based on the size of the target. A large vehicle may be out of range when the stadia lines show it to be in range. Second, the stadia lines are designed for a maximum engagement range of 3,000 meters.

(2) ***Binocular Method.*** Use the reticle in binoculars to determine if a target is within range by looking at the length, width, or height of the vehicle. Follow the same procedures when using the M17 and M19 binoculars, even though the reticles differ slightly. The M17 tick marks are only 1.7 mils long while the tick marks on the M19 reticle are 5 mils long (2.5 mils on each side of the horizontal and vertical scales).

(a) To determine if a target is within range at 3,000 meters based on the *length* of the target, place the length of the target on the vertical scale. If one-third or more of the vehicle extends beyond the tick mark, the vehicle is in range (Figure 5-4). A vehicle 6.5 meters long will measure about 2.2 mils at 3,000 meters and about 1.7 mils at 3,750 meters.

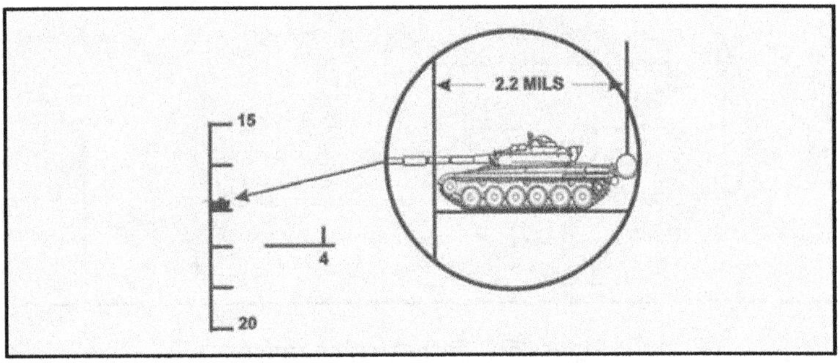

Figure 5-4. Target within range (flank or oblique) using M17 binoculars.

(b) To determine if a target is within range at 3,000 meters based on the *width* of the target, place the target on the small tick mark on the vertical scale. If the target covers two-thirds or more of the tick mark, the vehicle is within range (Figure 5-5, page 5-4). A vehicle 3.4 meters wide will measure 1.1 mils at 3,000 meters and .85 mils at 3,750 meters. (Most Warsaw Pact APCs are less than 3.4 meters wide and can be engaged at smaller mil values.)

(c) To determine if a target is within range at 3,000 meters based on the *height* of the target, place the target on one of the tick marks on the horizontal scale. If the height of the vehicle is one-half or more of the height of one of the tick marks, the vehicle is within range. A vehicle 2.4 meters high (the size of most Warsaw Pact vehicles), will measure .8 mils at 3,000 meters and .6 mils at 3,750 meters (Figure 5-6, page 5-4).

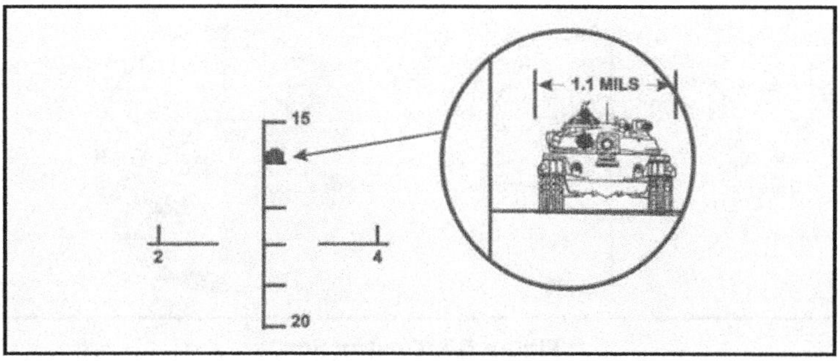

Figure 5-5. Target within range (frontal or rear) using M17 binoculars.

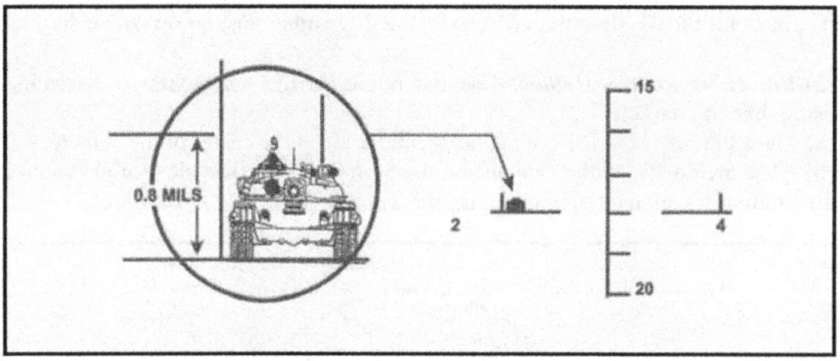

Figure 5-6. Target within range based on height.

NOTE: This method cannot be used if the weapon system is in an elevated firing position or if the lower portion of the target vehicle is hidden by foliage or terrain.

b. **Determine Exposure Time.** The half-sight method of determining exposure time is based on a vehicle speed of 35 kilometers per hour (the expected top vehicle speed of armored vehicles on level or gently sloping dry terrain).

(1) ***Daysight Tracker Method.*** Use the crosslines on the daysight tracker to determine the exposure time of a target.

(a) Place the crosslines of the daysight tracker on the center of the visible mass of the vehicle.

(b) If the area between the vertical crossline and the edge of the field of view in the direction of travel is clear of obstruction, the target is engageable (Figure 5-7).

(c) If obstructions appear between the vertical crossline and the edge of the field of view, the time of exposure would not be long enough for the missile flight before the target moved out of sight (Figure 5-7).

Figure 5-7. Obstruction.

(2) ***Nightsight Method.*** The procedure to determine exposure time is the same for the nightsight as for the daysight tracker, except the nightsight must be set on narrow field of view.

(3) ***Binocular Reticle Method.*** Use the reticle in the binoculars to determine the exposure time of a target.

(a) Place the zero tick mark of the horizontal scale at the center of the vehicle.

(b) If the area between the vehicle and the 50-mil tick mark is clear of obstructions in the direction of travel, the target is engageable (Figure 5-8).

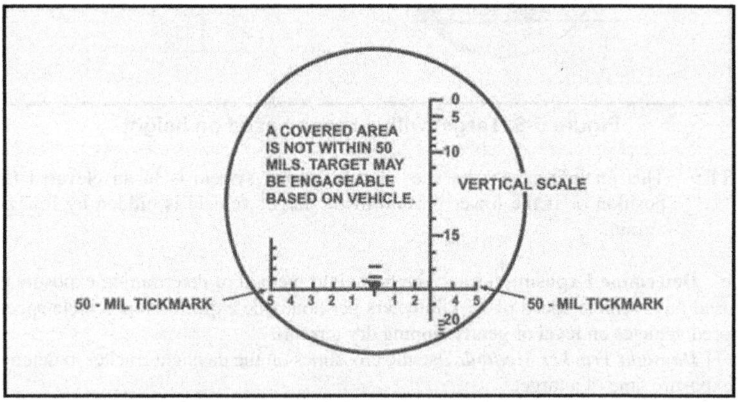

Figure 5-8. Binocular reticle method.

5-2. FIRE COMMANDS

The six elements of a fire command are alert, type of missile, target description, target direction, range, execution, and closing. Whether mounted or dismounted, the elements of the fire command are the same. (Figure 5-9, page 5-6, shows an example of a squad fire command.)

SQUAD LEADER:	SQUAD, TOW2A MISSILE; ONE TANK NORTH OF TRP 1; 2,000 METERS; AT MY COMMAND.
GUNNER:	IDENTIFIED.
SQUAD LEADER:	FIRE.
SQUAD LEADER:	CEASE TRACKING, OUT OF ACTION.

Figure 5-9. Example of a squad fire command.

a. **Alert.** The first element of the fire command alerts the crew for an immediate engagement. The squad leader commands, "Squad," and the gunner begins observing the target area.

b. **Type of Missile.** Because a variety of TOW missiles are used and their capabilities vary, a particular type of missile must be specified. A TOW crew in battle will probably have a mix of different missiles and a wide variety of target arrays to engage (see Chapter 1, TOW missile types and configurations). For example, if the target is a BTR-60 or BMP-1, the squad leader may command, TOW 2; if the target is a T-80 with reactive armor, he may command, TOW 2B.

c. **Target Description.** The second element identifies the target for the gunner. If several similar targets are present, this element tells the gunner which target to engage first. Most targets can be described by using the terms listed in Table 5-1. Targets that are combinations of the ones listed in Table 5-1, such as a truck mounting a missile system, are identified by combining terms—for example, TRUCK MOUNTING ANTITANK. When the gunner sees the target, he announces, "Identified." If multiple targets appear, the commander may specify which target will be engaged by the gunner--for example, FIRST TANK or RIGHT TRACK. (See Appendix H for information on OPFOR/Threat counterpart systems.)

TARGET	ANNOUNCED AS
Any tank or tank-like vehicle	**TANK**
Several tanks	**TANK FORMATION**
Any unarmored vehicle	**TRUCK**
Any halftrack or armored personnel carrier	**TRACK**
Helicopters	**CHOPPER**
All fixed-wing aircraft	**PLANE**
Personnel	**TROOPS**
Any machine gun	**MACHINE GUN**
Any antitank gun or towed artillery piece	**ANTITANK**
Any other target	**A short word or phrase that clearly describes the target.**

Table 5-1. Terms used to describe targets.

d. **Target Direction.** If the target is moving, the direction of movement is given after the description to aid the gunner in locating the target. After the gunner is given the location of a target, he can search for the target in the direction of movement. The following methods are also used to help the gunner locate the target.

(1) ***Target Reference Point.*** A TRP is an easily recognizable feature or point on the ground (either natural or man-made) used for identifying targets and controlling fires. They can be used to designate targets for companies, platoons, sections, and individual weapons. They can also be used to designate the center of an area where the commander plans to distribute or converge the fires of all his weapons. TRPs are usually designated by the company commander or platoon leaders.

(a) Weapons will engage targets from different directions, so compass points (for example, north, east), rather than "right" or "left," are used when giving directions centered on a TRP.

(b) Deflection from the TRP can be estimated, or it can be measured using the binocular mil scale or the circular reticle on the wide field of view–for example, 5.5 degrees at 1,000 meters is about 100 meters, or 200 meters at 2,000 meters.

(2) ***Prominent Features.*** The commander may give the distance and direction from a prominent feature–for example, FROM HILL SEVEN SIX TWO, LEFT TWO HUNDRED, or FROM BRIDGE, RIGHT FOUR HUNDRED.

e. **Range.** The range is given to help the gunner identify his target and to determine its engageability. The squad or section leader can determine the range to the target using the naked eye, binoculars (mil-relation formula), or reference materials (maps, range cards).

(1) ***Naked Eye.*** One method for using the naked eye to determine range is the football field method. The squad or section leader counts in 100-meter increments, estimating the number of football fields that could fit between the firing position and the target.

(2) ***Binoculars.*** Binoculars and the mil-relation formula can be used to determine range. To use this method, the squad or section leader must know the width, height, or length of the target. He determines the width, height, or length with the mil scale on the binoculars; substitutes the mil-relation; and computes the range.

(3) ***Reference Materials.*** Maps can be used to determine range by counting the grid lines between the firing position and the target or by adjusting from a known point. Range cards can also be used to determine the range to the target.

f. **Execution.** Two commands are necessary for execution: a preparatory command and a command of execution.

(1) AT MY COMMAND is a preparatory command that warns the gunner not to fire until given the command of execution.

(2) FIRE is the only command of execution used to fire a missile.

g. **Additional Commands.** In addition to the six elements of the fire command, some other commands are needed.

(1) The command CEASE TRACKING or CEASE TRACKING, OUT OF ACTION is issued after seeing the round detonate or when the squad or section leader wants to halt firing.

(a) CEASE TRACKING tells the crew the squad or section leader intends to stay in position and engage another target immediately or when one appears.

(b) CEASE TRACKING, OUT OF ACTION tells the crew the squad or section leader intends to move to another position.

(2) To determine the method of engagement, the section leader (or above) selects a fire pattern depending on the opposing force's formation. The section leader directs, FRONTAL, DEPTH, or CROSSFIRE, when the gunner is faced with multiple targets.

(3) When the target is identified, the gunner announces, "Identified."

(4) If the gunner cannot see the target, he announces, "Lost."

(5) If the gunner cannot identify the target, he announces, "Cannot identify."

(6) The loader announces, "Backblast clear," *before* the command of execution is given.

h. **Repeating Commands.** When a crew member fails to hear or understand any element of a fire command, he announces the element in question. For example, if the gunner asks, "Location?" the squad leader repeats the location element such as, "From hill seven six two, west two hundred."

i. **Correcting Errors.** To correct an error in a fire command, the squad leader announces, "Correction," and corrects only the element in error. He completes the command by announcing all elements after the corrected element. He does not try to correct an element that has been needlessly included, such as the direction element. He corrects the omission of an element by announcing "Correction" and then the omitted element. After announcing the omitted element, he completes the command.

j. **Commands for the Driver.** Although directions to the driver are not part of the fire command, they are given by the squad leader or gunner in short terms.

5-3. TARGET TRACKING

To track a target, the gunner visually acquires the target through the daysight tracker system of the TOW. He can track the target by either optical or electrooptical means, depending on the system configuration being used and on the visibility conditions.

a. To track the target, the gunner operates the hand controls on the traversing unit to keep the reticles in the launcher sight aligned with the target. The daysight tracker system is attached to and aligned with the launch tube. The launch tube stabilizes the exit of the missile from the launcher for initial alignment during missile flight. On achieving target alignment, the gunner fires the missile by manually depressing the trigger switch. Thereafter, all operations are automatic and the gunner's only task is to maintain alignment of the sight reticle on the target until missile impact.

b. Deviations of the missile from the line-of-sight trajectory are sensed in the launcher sight by infrared means that receive information from infrared radiators attached to the missile. This information is processed in the form of electrical signals to produce error signals proportional to the azimuth and elevation displacements of the missile from the intended trajectory. Correction commands are derived from these error signals and are sent to the missile over the command-link wires, which are dispensed from the missile. The missile performs corrective maneuvers using aerodynamic control surfaces that deflect in response to the command signals from the launcher. On target impact, a high-explosive, shaped-charge warhead is detonated.

5-4. TARGET ENGAGEMENT WITH THE M220A1 (BASIC TOW)

Specific procedures are followed to engage a target with the M220A1.

a. Position the eye well into the rubber eyepiece and place both hands firmly on the control knobs (Figure 5-10). Place the body so there is no contact between the shoulder and the encased missile. The only contact with the launcher is with the hands and eye. Assume a firing position that is comfortable. An uncomfortable position causes muscle tension, which affects the ability to track smoothly.

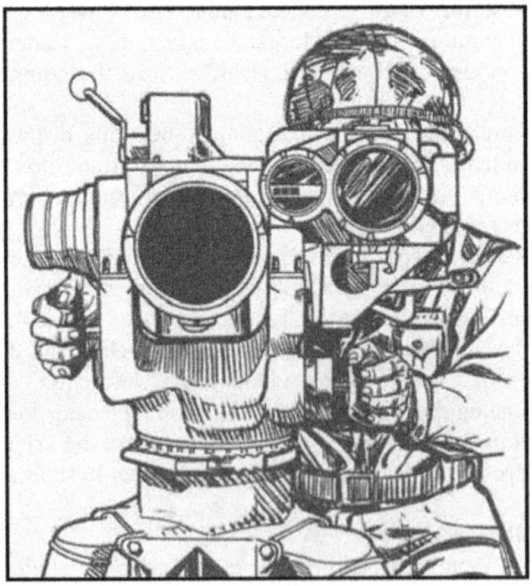

Figure 5-10. Control knobs.

NOTE: When firing from the tripod, kneel on one or both knees.

b. Raise the trigger protective cover and establish a smooth tracking rate while keeping the crosshairs on the center of visible mass of the target (Figure 5-11).

Figure 5-11. Crosshairs on the center of visible mass of the target.

c. Move the launch tube left or right by applying a smooth, steady force to both control knobs (pushing one and pulling on the other) and rotating the body from the waist up as the launcher moves. Elevate or depress the launch tube by applying a smooth, steady turning force to both control knobs. Applying pressure to only one control knob, or applying uneven pressure, makes it more difficult to track smoothly. Maintain the same arm, shoulder, and head position throughout an engagement. Any change in body position other than leaning with the controls will cause a jerking motion that could result in grounding of the missile.

d. Proper breath control is especially important during the first and last 400 meters of missile flight. Improper breathing will cause poor tracking. Take a deep breath and let part of it out; then, press the trigger. After a 1.5-second delay, the missile will launch. The delayed firing of the launch motor may cause you to flinch or jerk the control knobs if you are not prepared for it. Be prepared for two noises after the trigger is pressed. The first noise is the gyro being activated. While it is not loud, it may cause you to think a misfire has occurred, and you may not be prepared for the next noise. The second noise is the launch motor firing and it *is* loud. The dust, smoke, heat, and debris from the backblast may cause flinching. When the missile appears in the sight picture, ignore it. *Never try to guide the missile.* If distracted, tracking becomes poor and chances of hitting the target are reduced. Continue to track the target at a smooth tracking rate, keeping the crosshairs on the center of visible mass until missile impact (Figure 5-12).

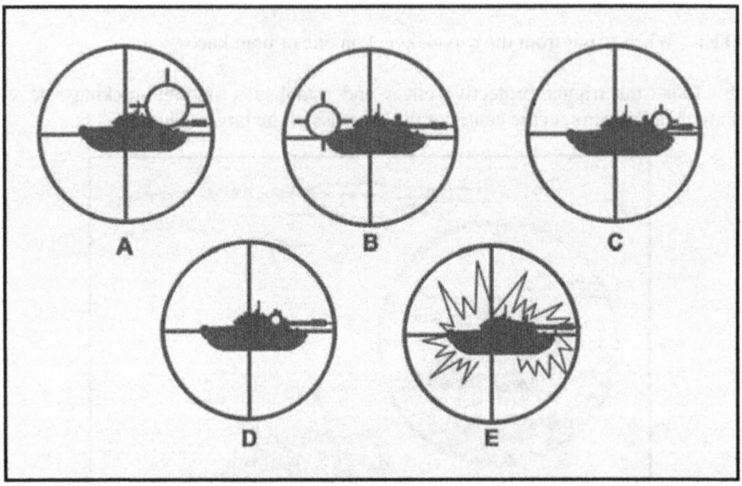

Figure 5-12. Keep crosshairs on the center of visible mass.

5-5. TARGET ENGAGEMENT WITH THE M220A2 (TOW 2)

Specific procedures are followed to engage a target using the M220A2.

a. Looking through the daysight tracker, adjust the focus control until the crosshairs are in focus. (To see the crosshairs clearly, set RETICLE LIGHT switch to ON.) Position the crosshairs on the target, and remove the front lens cover from the nightsight by releasing two latches. Set the ON-OFF-STBY switch to ON. Look through the eyepiece and adjust the diopter adjustment ring to focus the reticle. Ensure the battery monitor light is off. Set the field of view selector to wide field of view, locate the target, and adjust the range focus, contrast, and brightness controls to obtain the best possible sight picture. Set the field of view selector to narrow field of view and adjust the range focus, contrast, and brightness knobs.

(1) Because handoff may occur when limited visibility conditions exist, locate the target with the daysight tracker and fire using the nightsight. During daylight, the battlefield can be obscured by smoke, dust, and so forth. If limited visibility conditions exist, the MGS automatically transfers control of the missile from the daysight tracker to the nightsight. The nightsight allows continuous observation of the target regardless of battlefield visibility conditions.

(2) To view the battlefield, ensure the TOW 2 nightsight ON-OFF-STBY switch is in the ON position. Battlefield conditions requiring the nightsight to be in operation are area target monitoring and target engagement. The ON-OFF-STBY switch on the TOW 2 nightsight allows greater battery life from the BPC when placed in the STBY position. After the TOW 2 nightsight has been turned on and allowed to cool down (about three minutes), place the ON-OFF-STBY switch in the STBY position. The STBY position causes the closed-cycle cooler to cycle OFF for 100 seconds and ON for 20 seconds. Cool down is not required if the nightsight has been operating in the STBY position.

(3) When firing the TOW 2 system, ensure the nightsight is turned ON. Always allow three to five minutes cool down time before operation regardless of which sight is used. This ensures that the enhanced capabilities of TOW 2 are used during firing, flight, and impact. Certain conditions may permit degraded firing. During the system check-out procedure, if the display indicates a nightsight (postamplifier) failure, the TOW 2 can be fired using the daysight tracker only. If limited visibility causes handoff during flight, control of the missile may be lost.

b. To complete the firing sequence, follow the procedures for the basic TOW launcher (paragraph 5-4, page 5-9).

Section II. HELICOPTER ENGAGEMENT

Enemy armor is the primary threat to friendly ground forces employed in forward areas. The primary mission of the TOW is the destruction of these tanks at the greatest possible range. However, TOW gunners can also successfully engage attacking enemy helicopters, which are a significant threat to ground forces.

5-6. OPERATIONAL CONCEPT

Engaging helicopters with the TOW should be considered primarily as a means of self-defense. TOW crews should not consider helicopters as a routine target of opportunity, but they should leave them to conventional ADA assets when possible.

a. TOW positions are selected to cover armor avenues of approach, but these long-range fields of fire also facilitate the engagement of aircraft. The section leader's, squad leader's, and crew's observation from these positions can provide the early warning required to successfully engage aircraft.

b. The engagement of attacking helicopters should be done by TOW sections, not individual weapon systems. TOW sections should automatically engage helicopters that are attacking their positions. If one squad in a section is being attacked by a helicopter, the other squad should engage the helicopter while the first squad seeks cover. TOW crews and sections should be trained to automatically respond to helicopter attacks in this manner.

5-7. GUNNERY TRAINING

Specific gunnery training is required to track a helicopter with a TOW. Crews should perform the training according to the following conditions and standards. In addition to these outdoor training exercises, a number of scenarios for the TGT include helicopters as targets.

a. **Conditions.** Training takes place during daylight on a range. The TOW crew is given either a ground- or vehicle-mounted TOW with a TFTT and a helicopter mounted with TFTT reflectors. The helicopter moves toward the TOW gunner at speeds between 40 and 80 knots and at ranges between 1,000 and 3,500 meters. Three target angles are used for tracking: head-on; approaching at 30 degrees; and an evasive track with a 90-degree turn.

b. **Standards.** Within 5 seconds the gunner must acquire and begin tracking the target. He must get a hit 6 out of 10 times with the TFTT.

Section III. NBC AND LIMITED VISIBILITY CONDITIONS

Specific procedures are followed to operate the TOW during NBC and limited visibility conditions.

5-8. DECONTAMINATION

TOW crewmen must know decontamination procedures and materials. They must know which decontamination materials to use on each type of surface on the weapon system. Using the wrong material can cause damage to the system (for example, using DS2 on rubber surfaces). Decontamination materials and methods are periodically revised. (Refer to Appendix C of FM 3-5 for the most up-to-date information.)

5-9. THERMAL TARGET RECOGNITION, IDENTIFICATION, AND ENGAGEMENT

The nightsight allows the TOW gunner to view targets during limited visibility conditions such as darkness, smoke, fog, rain, and snow. It produces images called *thermal target signatures* or *infrared target signatures*, which are different from the images seen in the daysight tracker. Targets stand out in these infrared images and can be recognized at long ranges on a clear night and at reduced ranges during poor visibility. Recognizing these targets requires trained and experienced gunners.

a. **Temperature and Thermal Images.** Most objects have a radiated temperature either higher or lower than their background. Even if the radiated temperature differences are less than a degree, they appear on the nightsight display. If there is no difference between the temperature of an object and its background, the object will not be seen in the display.

(1) If an object has a high temperature, it will appear bright red in the nightsight. If the object has a low temperature, it will appear black. Usually, targets are easier to identify at night, because their radiated temperature is hotter than their background.

(2) Some targets, such as tanks and APCs, have internal temperature variations that form visible patterns. These patterns are the basis of target signature cues. In a nightsight, the shapes of the hottest vehicle parts, such as engines and exhausts, appear bright red. Objects with a medium temperature, such as the warm tracks, appear a dim red. Objects with a cool temperature, such as the cool hull and other cool parts, appear black.

b. **Sources of Infrared Energy.** Infrared energy comes from different sources such as solar heat, fuel combustion heat, frictional heat, and reflected radiance.

(1) ***Solar Heat.*** Solar heat comes from the sun and affects the exterior surface of objects. This heating highlights the outline of the object, which provides recognition cues to the gunner. These cues are usually similar to the overall appearance of the target. (For example, a solar-heated M113 appears box-like with a sloping front; a solar-heated M60 tank appears as a small oval atop a larger oval.) These shape cues are recognizable out to medium (1,000 to 2,500 meters) and long (beyond 2,500 meters) ranges. Since the sides have more defined contours, the side view shapes are usually easier to recognize than the front view. In addition to atmospheric variables and surface reflections, the solar heating rate is also affected by the object's ability to absorb sunlight. Generally, dark-colored objects are better absorbers of sunlight than light-colored objects.

(2) ***Fuel Combustion Heat.*** Fuel combustion heat comes from operating engines. The heat from operating engines is conducted to the surfaces of the surrounding engine compartment.

(a) Because engine compartment temperatures reach up to 200°F, the surfaces of these compartments radiate features that can be easily detected on the nightsight at long ranges. Heated personnel space is also visible.

(b) Engine muffler and exhaust pipe temperatures are high, providing the gunner with good cues.

(c) Although the engine, heated compartments, and exhaust features themselves do not appear in the nightsight, their cue value is not any less. A trained and experienced gunner can determine much about the vehicle from these cues.

(3) ***Frictional Heat.*** Frictional heat is produced by the moving parts of vehicles. However, these features usually appear a dim red. This heat is less intense than the high temperatures from the engine combustion. Frictional heat is generated only when the vehicle is in motion.

(a) Frictional heat provides long-range cues to classify the vehicle as wheeled or tracked. At medium-range to short-range, these cues can be used to identify the vehicle.

(b) The vehicle's transport systems are the source of most frictional heat cues. Tracked vehicles have frictional heat in the tracks, road wheels, drive sprockets, support rollers, and shock absorbers. The smallest of these features can be identified at longer ranges when they are hot. Wheeled vehicles have frictional heat in the tires, shock absorbers, drive shafts, transmissions, axles, and differentials. The tires, shock absorbers, and differentials can be detected at medium-range to long-range.

(4) ***Reflected Radiance.*** Certain smooth, glossy surfaces, such as windshields and glossy painted fenders, reflect radiation images from other sources. These reflections can produce odd images. For example, the fenders of a T-62 appear black because of thermal reflection. An overcast sky can cause warmer thermal reflections. Generally, surface reflections are diffuse in nature and do not usually cause problems.

5-10. EFFECTS OF WEATHER AND OBSCURANTS

Variations in solar heat, fuel combustion heat, frictional heat, and thermal reflection affect infrared signatures and infrared target recognition cues. In addition, some atmospheric conditions degrade the nightsight, while others can enhance it. Some of these factors are discussed below.

a. **Falling Precipitation.** Infrared energy does not transmit well through falling precipitation (rain, snow, and fog). The temperature of targets and background objects are decreased. The basic signature cues themselves do not change because of atmospheric transmission losses. Falling precipitation restricts nightsight visibility more than precipitation that has fallen.

(1) During rain or snow, background objects and frictionally heated and solar-heated target features lose heat. Frictional heat loss is caused by water and mud accumulating on the tracks, wheels, and other transport system parts. Engine compartment and exhaust temperatures remain high. Landmarks, such as tree lines, trails, and contour features, are often lost. The loss of heat in background objects reduces scene clutter, such as trees and rocks, and can increase target detection. Target recognition cues are usually reduced because of the loss of heat in certain target features.

(2) Because rain and snow have a cooling effect on the target's contrast, the nightsight contrast controls must be increased to compensate for the condition. However, a higher contrast setting produces a "snowy" image.

b. **Fallen Snow.** Fallen snow tends to make all ground temperatures the same. Depth perception by size comparison becomes difficult because of lack of terrain features with which to reference size.

c. **Dust, Diesel Fog, and Oil Smoke.** Dust particles from artillery impact greatly reduce nightsight visibility. Only the hotter objects and target features show through the obscurants.

5-11. COMPENSATION FOR TARGET APPEARANCE VARIABLES

Although vehicles have distinguishing characteristics or cues by which they can be classified and identified, vehicle appearance can be altered by changes in atmospheric and ground conditions. Therefore, the gunner must know to use the control settings of the nightsight to help compensate for these variables.

a. **Contrast and Brightness Control.** The contrast and brightness controls can be set for maximum internal detail of the target. The controls are balanced to give the clearest image of these target recognition cues. The following is general guidance for setting the image brightness and contrast controls. Gunners should be encouraged to experiment with the controls to understand the effects of the image controls on the thermal image and thermal signatures.

(1) ***Brightness, Low; Contrast, Medium to High.*** These settings are for scanning an area in search of targets. Background clutter is suppressed. Cool objects are not visible. Only the hot objects in the field of view are seen. When a possible target has been found, brightness can be increased and contrast can be lowered gradually to reveal more thermal detail in the suspected target. Low brightness and medium-to-high contrast settings are also for nightsight use in light fog, rain, or dust conditions.

(2) ***Brightness, Low to Medium; Contrast, Medium.*** This is the best overall setting for target detail. With medium contrast, brightness can be varied up and down to bring out features and determine the hottest vehicle parts. When brightness is lowered, the cooler parts, such as tracks, darken before hotter parts. Often, small changes in brightness can reveal much about the vehicle. For example, road wheels can sometimes be seen at long range if this technique is used. Experimentation is helpful in learning this target feature extraction technique.

(3) ***Brightness, Medium; Contrast, Medium to High.*** These settings work well in heavy fog or heavy dust when little can be seen with the TOW. The settings increase the snowy effect in the image, but they also increase the apparent sensitivity of the sight. The image appears distorted and is difficult to interpret. These settings are also good for searching a tree line in wet conditions. They are sometimes useful with long-range targets. Beyond 2,500 meters, small target images, such as the front view of a BRDM-2, will have a few recognizable features. At long ranges, the higher contrast setting highlights the vehicle's hull and overall silhouette. This will not provide internal detail, but it will help the target stand out from the background and will aid in target detection.

b. **Focus Controls.** Most nightsight focus controls are sensitive; that is, a small movement of the control knob results in a large change in focal point. If a gunner has difficulty in focusing, he should check the adjustment of the image controls.

(1) The eyepiece focus should be adjusted first. The eyepiece focus is called the *diopter adjustment*. Once a gunner knows his diopter correction number, he can dial it in on any nightsight diopter ring. Correct diopter adjustment can usually be obtained by focusing the eyepiece so that the reticle is focused.

(2) The second focus adjustment is the objective range focus. The image controls should be adjusted at the low-to-medium level before focusing the objective lens. Focusing the objective lens is learned through trial and error. It is made difficult by the fact that infrared heat diffuses on objects and does not usually give clear-cut, straight lines on which to focus. Thus, the gunner must learn to focus by adjusting the control back and forth to get the best image. This is easier to do when the nightsight is aimed at a prominent object. Once the best image is determined, the gunner can experiment by focusing on different objects at different ranges.

5-12. BATTLEFIELD IDENTIFICATION

Battlefield identification using a nightsight is difficult. Although the identification problem is being studied, little is known about the ranges at which high-confidence identification can be expected. In a target-rich environment on a dry, clear night, high-confidence identification requires a thermal image of such features as road wheels, turret shapes, gun tube, and exhaust location. Limited experience with tanks produced in the U.S. and other nations indicates that thermal identification can be made between 1,000 to 2,000 meters in clear weather. When identifying targets, a gunner should ask himself these questions:

- Is the target moving?
- What direction is it moving?
- Where is the engine?
- Where is the exhaust?
- Is the target in the unit's sector?
- Should it be there?
- Is it in a formation?
- Is it firing at the unit?

NOTE: Refer to FM 100-65 for primary recognition cues.

Section IV. ELECTROOPTICAL COUNTERMEASURES

TOW crews may encounter electrooptical countermeasures on the battlefield. Therefore, the following procedures should be used as engagement criteria and not as maintenance checks.

5-13. TOW LAUNCHER

TOW gunners should complete boresight procedures at least as often as prescribed in the operator's manual to ensure the system remains boresighted. When properly boresighted, the boresight meter needles are centered. The sight sensor must point at the ground or away from the sun or any other possible source of IR energy when the boresight procedure is being completed.

a. Once the boresight procedure is completed, the TOW crew leaves the self-test selector switch in position 7 and engages the test/operate switch (move to test position and hold) as the gunner acquires the target. The missile arming lever should not be raised at this time. If the self-test meter needles peg to one side, or become erratic, or both as the gunner aims at the target, electro-optical countermeasures are present and a shot against that target should not be attempted. The gunner waits for a flank shot or shifts targets.

b. The loader releases the test/operate switch as the gunner shifts targets. The loader reengages the test/operate switch (returns it to test position) when the gunner is ready to acquire a second target. Releasing the switch saves wear and tear on the daysight tracker sensor and conserves battery power, if batteries are being used.

c. If the meter needles return to center after the gunner places his crosshairs on a second target, he engages the target. The loader releases the test/operate switch and raises the arming lever, and the gunner fires the missile.

d. If the meter's needles are erratic after the gunner lays on the second target, he does not fire. He selects another target or waits for an oblique or flank shot ensuring the meter needles are centered before firing.

5-14. TOW 2 LAUNCHER

TOW 2 launchers firing TOW 2-series missiles are not vulnerable to electrooptical countermeasures. However, TOW 2 launchers firing non-TOW 2 missiles are vulnerable, and gunners should use the boresight indicators to determine whether a target should be engaged.

a. The loader completes the MGS self-test procedure as target engagement becomes imminent. The loader holds the self-test toggle switch in the test position before the missile is armed, as the gunner places his crosshairs on the target. If the green light stays on indicating electro-optical countermeasures are not present, the loader releases the toggle switch and raises the arming lever, and the gunner engages the target.

b. If the green boresight center light goes off and the red lights begin to flicker on and off indicating electro-optical countermeasures are present, the gunner does not engage the target. The loader holds the self-test toggle switch in the test position as the gunner acquires an alternate target. If the red lights go off and the green light comes on as the gunner lays his crosshairs on a second target, the loader releases the self-test toggle switch and arms the missile, and the gunner engages the target.

c. Because the TOW 2 self-test switch is reactivated each time the toggle switch is placed in the test position and the self-test cycle takes 15 to 20 seconds, the loader does not release the test switch while the gunner is acquiring alternate targets. The leader releases the switch only after the green boresight light appears, signifying that it is safe to engage that target.

APPENDIX A

TOW TRAINING TIPS

Experiences at the National Training Center have produced many useful tips for training and employing TOWs. Following these tips will greatly enhance the performance of TOW units in the field.

1. To reduce the effects of heat scintillation, fire from a higher to a lower elevation whenever possible.

2. Train crewmen to remove the protective covers from the electrical connectors of missiles before loading. This is the most common cause of misfires during live fire.

3. Emphasize the obstacle effects of TOW guidance wires. Wire will impede or stop vehicular and dismounted movement. Plan for their recovery and or destruction. Wire cutters are not fine enough to cut TOW wire. The wire must be cut with low-intensity flame when possible.

4. Perform a system self-test and collimate the nightsight to the daysight tracker every 4 hours, after every 10-degree change in temperature; and after moving cross country.

5. Select flat, level firing positions to improve hit probability.

6. Practice fire control and fire patterns at every opportunity.

7. Identify maximum engagement lines and TRPs as accurately as possible. Use adjacent tanks or FO/FIST laser range finders when available.

8. Perform the 180-day verification checks as often as possible (every 30 days is not too often). Verify in the assembly area immediately before combat or live-fire operations. Allow time to fix deficiencies.

9. Increase the amount of time spent tracking in MOPP4. Tracking in MOPP4 improves tracking skills even under MOPP1 conditions.

10. Ensure TOW live fire is conducted from the narrow field of view only.

11. Ensure M966s and BFVs with TOW 2 weapon systems receive priority issue of TOW 2 missiles.

12. Collimate nightsight to daysight each time the system is boresighted.

13. Know the diopter setting. This information should be posted on the system for ready reference.

14. M966 TOW HMMWV:

a. Shield the nightsight and daysight from direct sunlight. Cool the nightsight as much as possible to reduce system cool-down time.

b. Familiarize gunners with launch obscuration effects:

(1) *White out*—1.5 seconds after trigger pull, block daysight for 2 to 5 seconds.

(2) *Red out*—1.5 seconds after trigger pull, pass a lit cigarette lighter or other IR heat source 6 inches in front of the AN/TAS-4.

c. Remove AN/TAS-4 eyepiece when tracking in a protective mask. Replace with a doughnut-shaped foam cushion made from a sleeping mat or from ammunition packing and taped to AN/TAS-4.

d. Reinforce counter-rotation launch technique to reduce launch tube climb during live fire.

e. Train gunners to bring crosshairs back on target slowly, avoiding jerky movements.

f. Minimize movement of crew members on vehicle when missile is in flight.

g. Minimize blowing of sand and dust from missile launch, by placing a tarp on the ground in front of the launch tube or by wetting the ground down. Blowing dust and sand can cause the weapon system to lose the IR beacon on the missile.

15. Fire Commands/Missile Selection. With the increase in the number of different types of missiles in the TOW inventory, it will sometimes be necessary to tell the gunner what type of missile to engage the target with (see Appendix H, Table H-2). If a missile is not loaded in the weapon system, the squad leader can include the missile type in the target description portion of the fire command (SQUAD, T-80, TOW 2B). If a missile is loaded in the weapon system, the squad leader will have to make the decision whether time and the tactical situation make it desirable to unload and then reload the system. If a missile has just been fired, the missile reload type can be designated as part of the cease tracking command (CEASE TRACKING, TARGET DESTROYED, RELOAD WITH TOW 2A).

16. TOW MILES Gunnery. Because TOW MILES is an engagement simulator adapted as an interim gunnery trainer, it is important to return gunners to a precision tracking system before live-fire or combat operations. Because the tracking simulation of TOW MILES is less realistic and more forgiving of errors, hit success with the MILES simulator comes more easily than hit success with the actual TOW missile. This is especially true with panel targets equipped with the LTID. Exclusive training on TOW MILES sometimes lulls gunners into a false sense of tracking proficiency. As an engagement simulator in the role of tactical trainer, TOW MILES offers adequate tactical realism. Good TOW gunners are usually good MILES gunners. The reverse is not necessarily true.

17. TOW Missile Clearance Requirements. If line-of-sight clearance is less than 30 inches, the probability of the missile hitting the ground or an obstruction is increased. Figure A-1 shows the probability of survival for the TOW. The reason for the deviation is that the missile does not precisely follow a gunner's line of sight.

RANGE TO TARGET	HEIGHT OF LINE OF SIGHT ABOVE THE GROUND: 18 INCHES	20 INCHES	30 INCHES
200M	.98	.98	1.0
300M	.95	.95	1.0
400M	.91	.91	.98
500M	.86	.91	.98
1,000 TO 3,000/3,750M	.56	.61	.93

Figure A-1. Probability of survival for the TOW (meters).

18. TOW Missile Time-Of-Flight Versus Range. Figure A-2 (page A-4) and Figure A-3 (page A-5) show the time of flight of the TOW missile to various ranges. Using these figures allows the squad leader to accurately calculate the time it takes for the missile to reach its target.

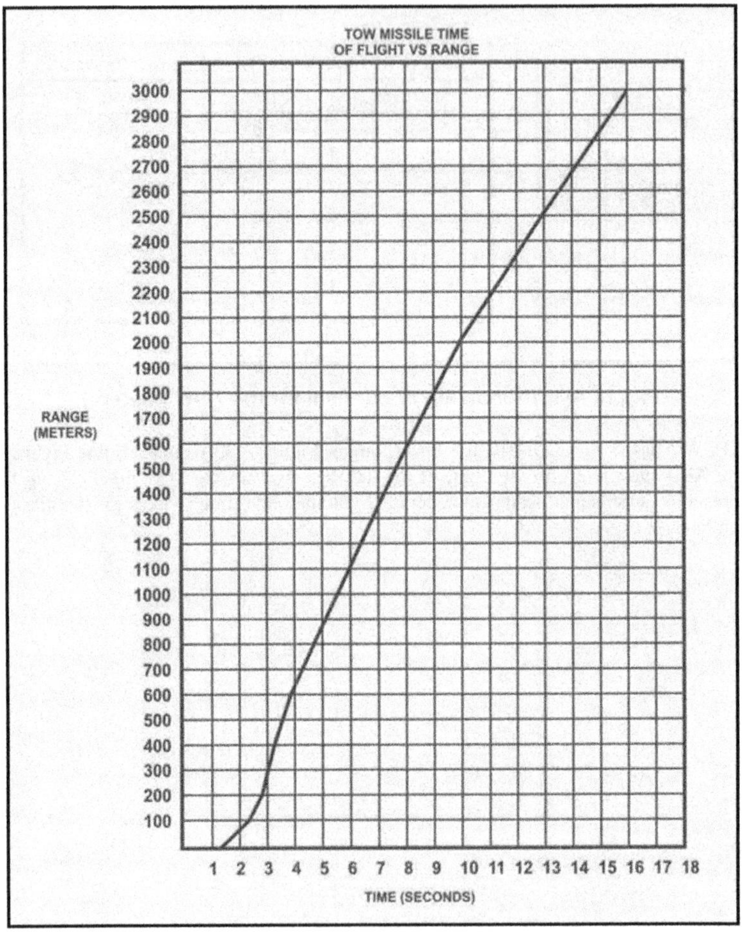

Figure A-2. TOW missile time-of-flight versus range.

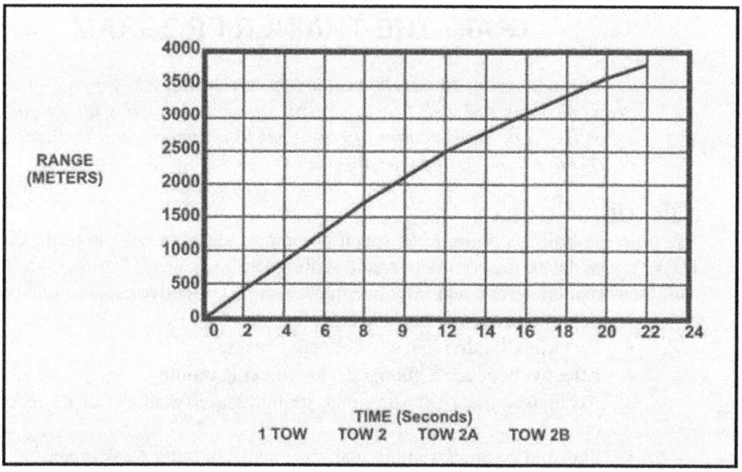

Figure A-3. TOW missile time-of-flight versus range (kilometers).

APPENDIX B

TRAIN-THE-TRAINER PROGRAM

The success of the TOW training program depends on trainers who are competent and well prepared. This appendix provides information to assist the TOW unit's chain of command in preparing and implementing an effective train-the-trainer program.

B-1. OBJECTIVES

The train-the-trainer program has specific objectives. Its purpose is to develop in every TOW trainer the skills, confidence, and willingness to train 11B soldiers to be proficient with their weapon system and effective in combat. An effective trainer must be able to—

- Set up and operate training equipment.
- Perform all individual and collective tasks.
- Effectively coach a gunner during tracking training.
- Accurately assess the training strengths and weaknesses of individuals and crews.
- Plan and conduct training that will correct training weaknesses.
- Instill confidence in the individuals and crews he trains.

B-2. MISSION-ESSENTIAL TASK LIST

Each commander of a TOW unit should carefully examine his unit's wartime mission and develop a METL. He then organizes a unit sustainment training program that supports that METL. The TOW Training Program consists of quarterly and semiannual mandatory training events. That does not mean that no TOW training should be done in the interval between these events. The commander is responsible for assessing his wartime mission and his unit's current training status and then devising a unit sustainment training program. That program should both support his METL and prepare his unit for the next round of TOW gunnery tables.

B-3. TRAINER ASSESSMENT

Trainers in a TOW unit are normally the squad and section leaders and the platoon sergeants. The trainers are not chosen by their proficiency but by their duty position. Therefore, assessing the abilities of all element leaders is critical. Any shortcomings must be corrected. This assessment is primarily the responsibility of the commander and the platoon leaders and sergeants. This can be accomplished in several ways.

- Conduct periodic leader's tests, both written and hands on.
- Have all leaders perform all the tasks that their subordinates are required to take in the quarterly and semiannual mandatory training events (Gunner's Skill Test and TOW Gunnery Tables).
- Hold a competitive skill test each month with the winner being named trainer of the month.

Any of these methods, or any other the commander deems appropriate, are acceptable as long as they provide an accurate assessment. Trainers should be thoroughly refreshed in the skills they are deficient in before they are allowed to train and test subordinates. A soldier must know his superiors are more skilled than he is in order to have confidence in those superiors.

B-4. COMMAND BENEFITS

If a train-the-trainer program is conducted aggressively and regularly, the benefits to the unit are considerable.

a. The base of trainer expertise will remain solid and gradually spread through the unit.

b. Maintaining this base is easier than periodically recreating it.

c. Overall unit proficiency and esprit de corps will steadily rise.

d. A well-trained and proficient unit requires less time for sustainment training and leaves more time for other unit missions.

B-5. TRAINING TASKS

To be successful, the trainer must be thoroughly familiar with the operation of his training devices. All trainers must be able to correctly perform the following tasks:

a. **TOW Gunnery Trainer.**

TASK: Assemble the TGT.
CONDITIONS: Indoors, given a TOW tripod, traversing unit, and launch tube and all components of the TGT.
STANDARDS: Correctly assemble all components of the TGT IAW TM 9-6920-452-10.

TASK: Prepare the TGT for operation.
CONDITIONS: Indoors, given a fully assembled and operational TGT.
STANDARDS: Correctly power on and warm up the TGT, and perform the operational checkout of the TGT IAW TM 9-6920-452-10.

TASK: Conduct pre-operational procedures for the TGT.
CONDITIONS: Indoors, given an operational and fully assembled TGT.
STANDARDS: Correctly load the videodisc, program disk, and the data disk IAW TM 9-6920-452-10.

TASK: Edit a planned group.
CONDITIONS: Indoors, given a fully assembled, operational TGT and matching data and videodiscs.
STANDARDS: Correctly edit a planned group to match the desired conditions in the mission IAW TM 9-6920-452-10.

TASK: Perform operational procedures for the TGT.
CONDITIONS: Indoors, given an assembled, fully operational TGT with matching video and data disks and an edited planned group.
STANDARDS: Correctly brief the student and conduct a mission from the planned group IAW TM 9-6920-452-10.

---

b. **MILES.**
TASK: Install MILES on the TOW weapon system.
CONDITIONS: Outdoors, given a TOW traversing unit, tripod, nightsight, and launch tube and a TOW MILES tracker head, ATWESS tube, and MGS simulator.
STANDARDS: Correctly assemble, install, and test the TOW weapon system with TOW MILES IAW TM 9-1265-368-10-2 (tasks 9 through 12).

TASK: Test the assembled TOW MILES.
CONDITIONS: Outdoors, given an assembled TOW weapon system with TOW MILES installed.
STANDARDS: Correctly test the TOW MILES IAW TM 9-1265-368-10-2 (task 13).

TASK: Fire TOW and ATWESS.
CONDITIONS: Outdoors, given an assembled, tested, and operational TOW MILES system.
STANDARDS: Correctly and safely fire TOW MILES with ATWESS IAW TM 9-1265-368-10-2 (task 15).

TASK: Reset MWLD and TOW MILES.
CONDITIONS: Outdoors, given an MWLD, an assembled, tested, and operational TOW MILES system, and a controller key.
STANDARDS: Correctly reset the MWLD and TOW MILES after they have been "killed" IAW TM 9-1265-368-10-2 17).

---

c. **TOW Field Tactical Trainer.**
TASK: Assemble the TFTT and install the retroreflector.
CONDITIONS: Outdoors, given a complete TOW system and a complete TFTT with retroreflector.
STANDARDS: Correctly install the TFTT and retroreflector IAW TM 9-6920-453-10 (M966).

TASK: Power up the TFTT and align the laser to the retroreflector.
CONDITIONS: Outdoors, given an assembled TFTT and an installed retroreflector.
STANDARDS: Correctly power up the TFTT and align the laser to the retroreflector IAW TM 9-6920-453-10 (M966).

TASK: Conduct operational procedures for the TFTT.
CONDITIONS: Outdoors, given an assembled, operational, aligned TFTT, and an installed retroreflector.
STANDARDS: Correctly conduct operational procedures for the TFTT to included setting target size and obscuration time, loading the M80 blast simulator, and preparing the TFTT for firing IAW TM 9-6920-453-10 (M966).

TASK: Conduct system checks and test procedures for the TFTT.
CONDITIONS: Outdoors, given an assembled, operational, and aligned TFTT.
STANDARDS: Correctly perform all system checks and test procedures for the TFTT to include visual checks, cable checks, power checks, computer checks, alignment checks, and system and operational checks IAW TM 9-6920-453-10 (M966).

B-6. TRAINER CERTIFICATION PROGRAM

The commander may choose whatever method he thinks best to assess trainers, but he must include the following elements in a trainer certification program:

a. **Program Orientation.** When a new trainer first enters the unit, he must be briefed on the unit's instructor certification program so that he understands what is expected of him.

b. **Testing of 11B Individual Soldier Tasks.** Each trainer must be tested on each individual task at least annually. Testing every trainer on every task more frequently is impractical. A more workable method is to conduct monthly or quarterly testing events where some portion of these tasks are tested, and over the course of a year all tasks are included.

c. **Testing of Collective Tasks.** Trainers should be tested in all collective tasks annually. Testing events should be spread out in partial monthly or quarterly events.

d. **Testing of Training Equipment Tasks.** This should be conducted in the same manner as the individual and collective tasks. The testable training equipment tasks are listed in paragraph B-5. These tasks are not found in the soldier's manual. The appropriate TMs must be referenced for the performance measures.

e. **Formal Records Keeping.** Formal records should be carefully maintained. If a trainer fails a particular task, he should be retrained and retested until he receives a go on it and all other tested tasks. No one should be allowed to train troops who has not received a go on all tasks tested to date. The soldiers' full confidence in the abilities of their trainers and leaders is critical to morale.

APPENDIX C

FIGHTING POSITIONS

TOW fighting positions may be mounted or dismounted, depending on the situation. A camouflage net can be used with either configuration.

C-1. DISMOUNTED TOW FIGHTING POSITION

A tripod-mounted TOW fighting position must meet the following specifications:

- Allows the squad to engage the enemy in its assigned sector of fire.
- Provides protection for both the weapon and its crew.
- Blends with its surroundings so that it cannot easily be detected 35 meters to the front and cannot be seen from the air.

a. The steps in constructing the position should be conducted in the following sequence:

(1) Assemble the launcher.

(2) Clear the fields of fire while ensuring the sector is under observation.

(3) Dig the weapon position first, then add overhead protection for the crew.

(4) Build a parapet to the front and flanks of the position.

(5) Disconnect the MGS and place it in the position made for it. Place the launcher into the position. Reconnect the MGS and check the boresight.

(6) Improve the position by adding overhead cover for the crew and system.

(7) Camouflage the position.

(8) Inspect the position and its camouflage. If possible, move at least 35 meters to the front of the position and study it.

b. The initial position should look like Figure C-1.

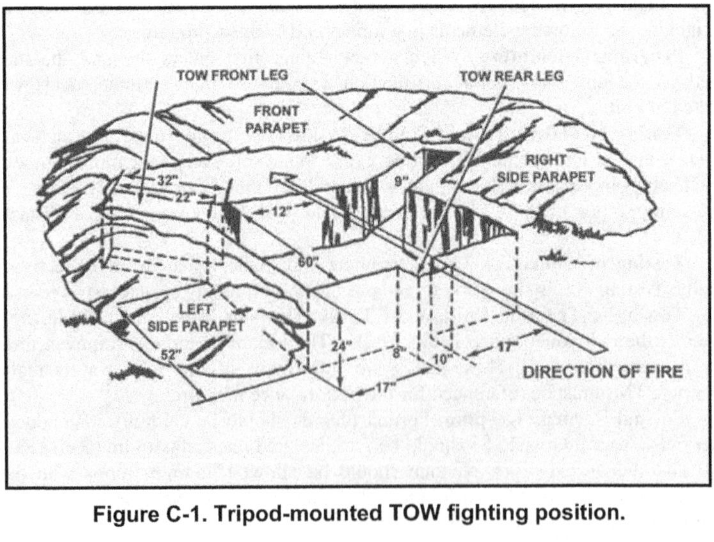

Figure C-1. Tripod-mounted TOW fighting position.

(1) Make the position 24 inches deep.

(2) Make the parapet at least 18 inches thick to keep out small-arms fire and artillery fragments. It should provide 9 inches of muzzle clearance under the launch tube.

(3) Do not place dirt or equipment in the backblast area.

(4) Scoop out a place for the MGS either under or to the front of the tripod.

(5) Dig a storage/protective area for the crew and missiles to one flank at a 90-degree angle to the primary direction of fire. Use the strongest material available for the roof. Put canvas or plastic down before throwing dirt on the roof to keep the ceiling from leaking. Place at least 20 inches of dirt on top of the storage/protective area (Figure C-2).

(6) Ensure the ground behind the TOW is free of leaves and dirt out to 25 meters so the backblast does not leave a signature.

(7) Ensure all the standard principles of camouflage are followed. (For example, cover all fresh dirt with leaves and brush, replace withered foliage, always approach position from rear, and so forth.)

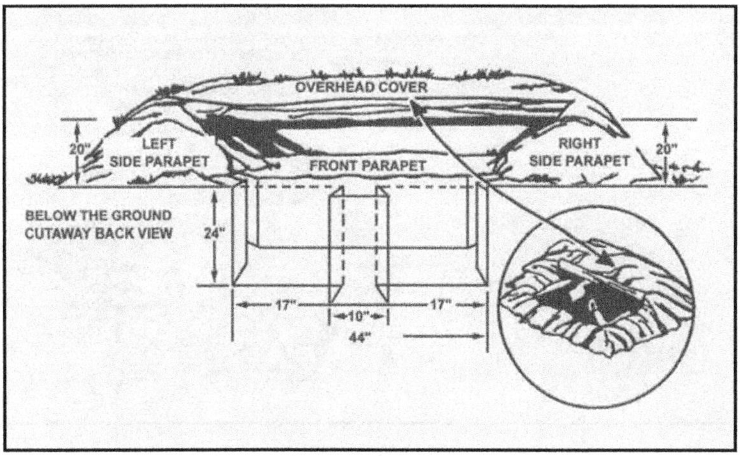

Figure C-2. Storage/protective area.

C-2. MOUNTED TOW FIGHTING POSITION

A mounted TOW fighting position must meet the following specifications:

- Does not restrict the TOW's target engagement.
- Provides concealment from ground and air observation for the crew and vehicle.
- Backblast area is as clear as possible of debris that would increase launch signature.

a. Camouflage and conceal the firing position.

(1) All vehicles travel to a position over the same route to prevent the enemy from detecting the number of vehicles present.

(2) The route taken should follow existing paths, roads, fences, or natural lines in terrain.

(3) Exposed routes should not end at a position.

(4) If only a short portion of the route into a position is exposed, sweep out the tracks with tree branches.

(5) Traffic in and out of a position should be held to essential movements only.

(6) Upon moving into a position, be sure to erase the trail leading into the position.

(7) Ensure all standard principles of camouflage are adhered to (same as for tripod-mounted position).

b. If a hull-down position is not available, use a hide position (Figure C-3).

c. If possible, move at least 35 meters in front of the position to ensure that it looks natural and blends with its surroundings.

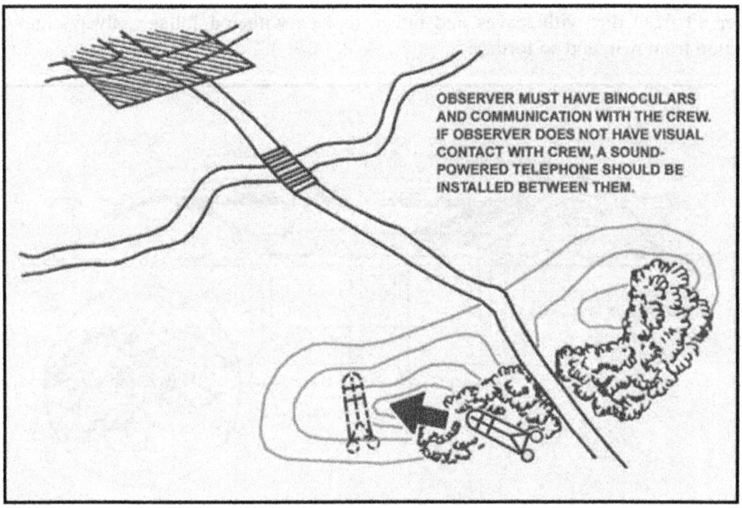

Figure C-3. TOW hide position.

C-3. CAMOUFLAGE NETS

When camouflaging any TOW fighting position, whether mounted or dismounted, the net must be positioned so it does not hang in the backblast area. If any portion of the net is in the backblast area, the net will be damaged and could collapse on top of the launcher when a missile is launched. Also, the net must not interfere with the field of fire.

APPENDIX D
STANDARD RANGE CARD

This appendix describes the various parts of a standard range card. It also provides a detailed discussion of how to complete a range card.

D-1. DESCRIPTION

A range card is a sketch of the terrain a weapon system has been assigned to cover by fire. It contains information that assists in the planning and controlling of fires, the rapid detection and engagement of targets, and the orientation of replacement personnel or units. By using a range card, a gunner can quickly and accurately determine the information he needs to engage targets.

a. A sector of fire is a part of the battlefield within which a gunner is responsible for engaging targets. Sectors of fire are assigned to ensure weapon systems will cover target approaches. Leaders should strive to overlap sectors to cover areas that cannot be engaged by one system. The leader gives a gunner boundaries running between prominent terrain features, or by left and right limits indicated by terrain features or azimuths. If necessary, the leader also assigns a gunner more than one sector of fire, designating one sector as primary and others as secondary.

b. The section or squad leader may also designate anticipated target engagement locations within the sector of fire. Those are recognizable terrain features on or near likely enemy avenues of approach. This information is placed on the range card.

c. Leaders may pick out natural or man-made terrain features that can be used as reference points for locating targets and adjusting direct and or indirect fires. Those features are called target reference points (TRPs). TRPs are requested through the mortar or artillery fire support team (FIST) or fire support officer (FSO). If TRPs are in or near the sector of fire, the leader should point them out and tell the gunner their numbers. If he does, the gunner depicts the TRPs on his range card. Normally, a gunner has at least one TRP, but not more than three, in his sector of fire.

d. Natural or man-made terrain features, such as hills, draws, or buildings, may be within the sector(s) of fire that prevent the gunner from firing in that area. The area blocked by these features is called dead space. All dead space in the sector(s) of fire must be determined so leaders can plan other weapon systems or other types of fire to cover the area (for example, mortars, artillery, or mines). Dead space is indicated on the range card.

e. The length of the sector of fire is normally limited by the maximum engagement range of the antiarmor weapon, but it can be less if any natural or man-made terrain features (trees, fences) prevent the gunner from engaging targets at maximum range. Regardless of what affects it, the maximum engagement range is shown on the card as a maximum engagement line. The squad leader uses a map to determine the distance to the maximum engagement line.

f. All TRPs, anticipated target engagement areas, azimuth and distance to a known point, and left and right limits are numbered on the sector sketch and in the data section with corresponding numbers for quick response.

D-2. PREPARATION PROCEDURES

The section/squad leader's briefing provides the necessary information for the gunner to begin preparing DA Form 5517-R (Standard Range Card). If he is assigned alternate and supplementary firing positions, he prepares a range card for them also. The gunner prepares the range card by performing the following steps.

EXAMPLE OF SECTION OR SQUAD LEADER'S BRIEFING:
"Our mission is to cover a sector of fire that begins at our present position and goes in the direction of the windmill to the maximum engagement range of 3,750 meters; it extends to the right across the high ground behind the houses, and hill, to the right edge of the orchard and returns here. The enemy should approach from the north and will probably use both Marshall Road and Lewis Road to enter our sector. We must plan on engaging the enemy in this area as soon as he is within range. There are two target reference points within your sector; the road intersection of Marshall Road and Duffell Road is TRP-Charlie One and the road junction of Lewis Road and Duffell Road is TRP-Charlie Two. Use the road junction of Campbell Road and Lewis Road to your left as a reference point to locate your position. The distance from the road junction is 633 meters on an azimuth of 85 degrees." (See Figure D-1.)

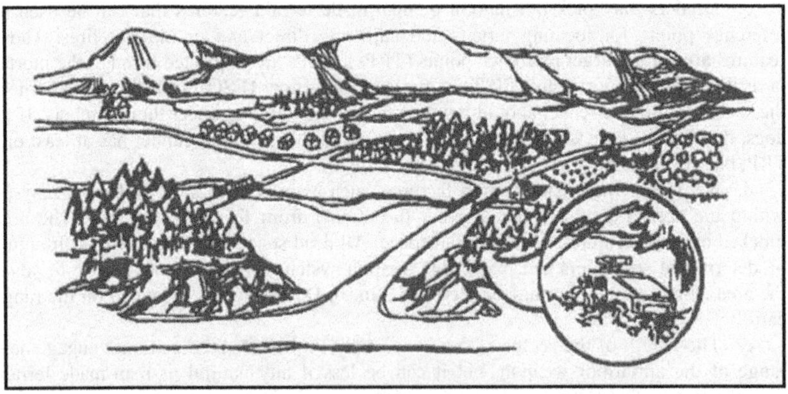

Figure D-1. Section or squad leader's briefing.

a. Draw a sector sketch of the entire sector. Make the sketch as large as possible, not to exceed the largest circle. For a large area covered by trees or woods, draw only the outline and label the area; for example, "woods" or "orchard" (Figure D-2).

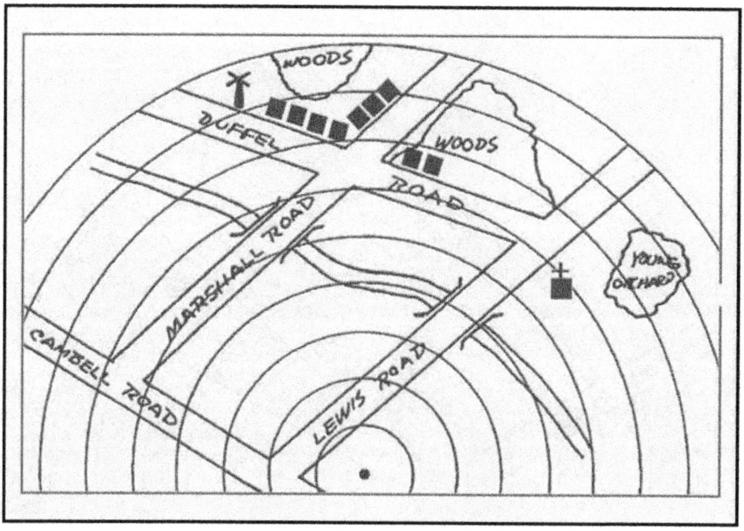

Figure D-2. Sketch of area on range card.

b. Draw lines from the weapon position (indicated by the black dot at lower center of range card) to show the right and left limits. Place a number 1 at the end of the left limit line and draw a circle around the number. Place a number 2 at the end of the right limit line and draw a circle around the number (Figure D-3, page D-4).

NOTE: After drawing left and right limit lines, the weapon symbol can be drawn over the black dot.

c. If there are no limitations, the maximum engagement line is curved and joins the left and right sector of fire boundaries at the maximum engagement range (Figure D-4, page D-4). If there are limitations, the maximum engagement line is drawn in front of the limiting terrain feature.

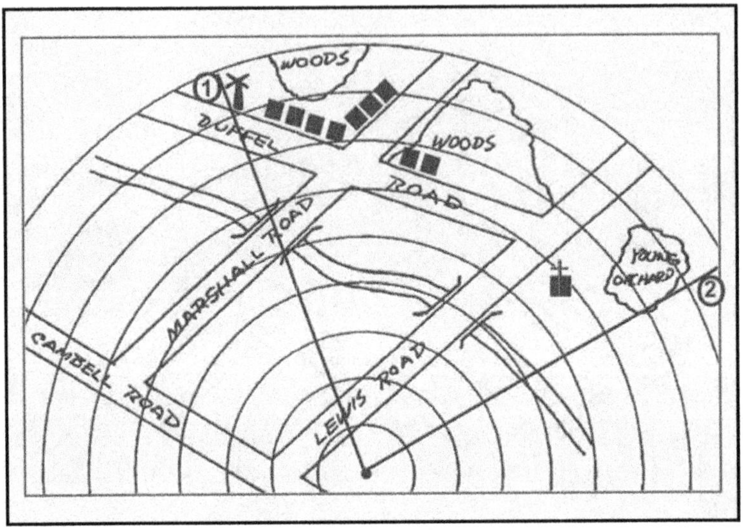

Figure D-3. Labeling sector of fire.

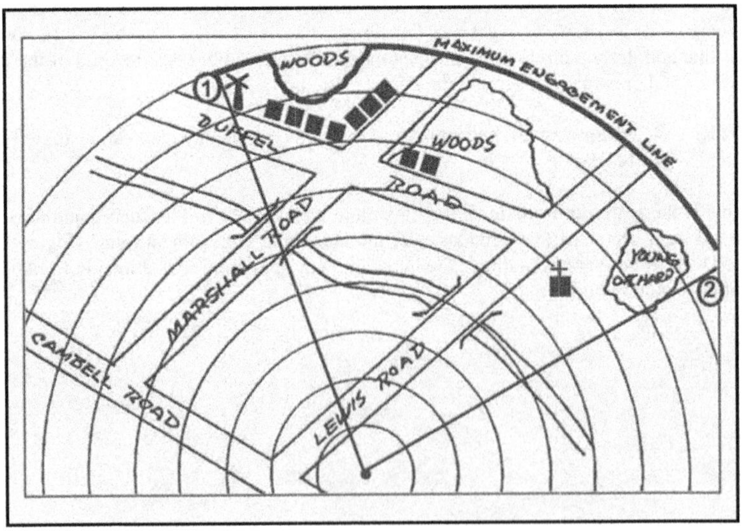

Figure D-4. Placement of maximum engagement line.

d. Number the anticipated target engagement areas (ATEAs) from left to right, starting with number 3. Place a number at the maximum engagement range of the target on the range card and circle the number (Figure D-5, page D-6).

e. Number the TRPs from left to right. Place the number below or next to the TRP on the range card and circle the number (Figure D-6, page D-6).

f. Place diagonal lines, or the words "dead space," where dead space occurs (Figure D-7, page D-7).

g. Use a compass to determine the azimuth from the firing position to the known point. Convert the direction to a back azimuth. Draw a line with multiple arrows from the known point to the firing position. Place a number at the known point and circle the number (Figure D-8, page D-7).

h. Fill in the marginal information at the top of the card.

(1) ***Unit Description–SQD, PLT, CO.*** Never indicate a unit higher than company level.

(2) ***Magnetic North.*** Orient the range card with the terrain. Place the compass on the range card. Determine the direction of magnetic north arrow and mark it on the card.

i. Fill in the data section at the bottom of the card.

(1) ***Position Identification.*** List either primary, alternate, or supplementary.

(2) ***Weapon***. See Figure D-9 (page D-8) for weapon symbols.

(3) ***Date***. List the day and month.

(4) ***Each circle equals ______ meters.*** Write the distance between the circles in meters. To determine the distance, count the intervals from the weapon to the maximum engagement line (as determined by the squad leader). Divide the amount of intervals into the range of the maximum engagement line. This will give the distance between circles (Figure D-10, page D-8).

Example: 9 intervals into 3,750 meters = 416 meters between circles.

(5) ***No. (Number)***. Starting with number 1, list the left limit, the right limit, and locations of ATEAs and TRPs shown on the sector.

(6) ***Direction/Deflection.*** Only degrees or the azimuth from the azimuth bevel ring (improved TOW vehicle) is listed. Line through the word that does not describe the information listed.

(7) ***Elevation***. This is only used with a ground-mounted machine gun using the traverse and elevation mechanism.

(8) ***Range***. Distance in meters from the weapon to the TRP or target engagement area.

(9) ***Ammunition***. List the type of ammunition used, if applicable.

(10) ***Description***. List the name of the object; for example, road, windmill, church. If the item is a TRP, also list the TRP number.

(11) ***Remarks***. Enter the weapon's reference point and any additional information not listed in the range card section. If more space in the data section is needed, use the reverse side of the range card.

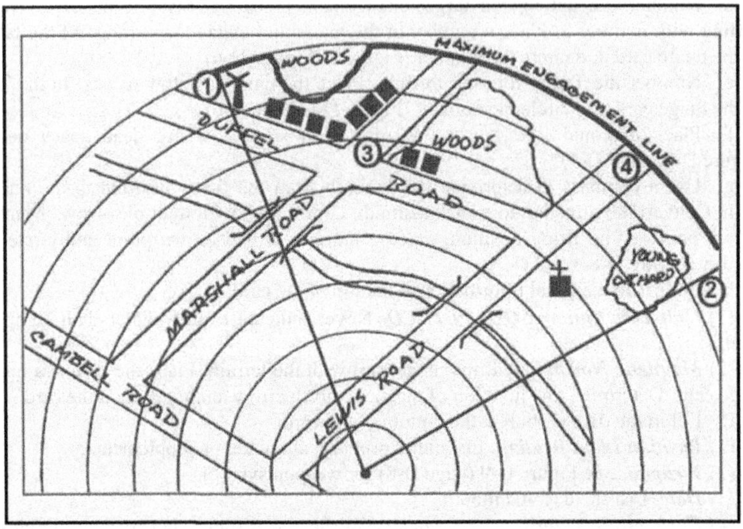

Figure D-5. Numbered target engagement locations.

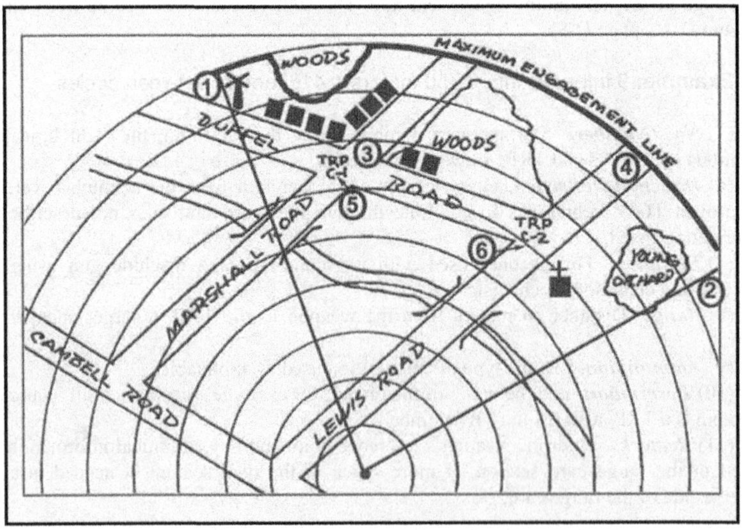

Figure D-6. Target reference points and numbers.

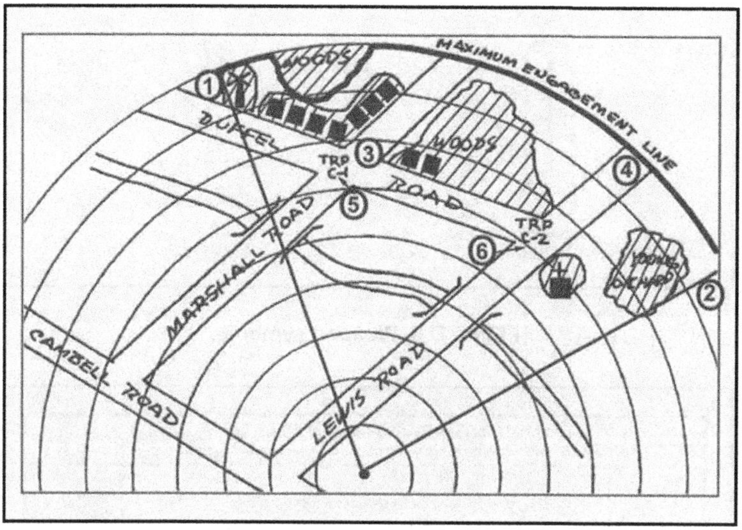

Figure D-7. Dead space indicated.

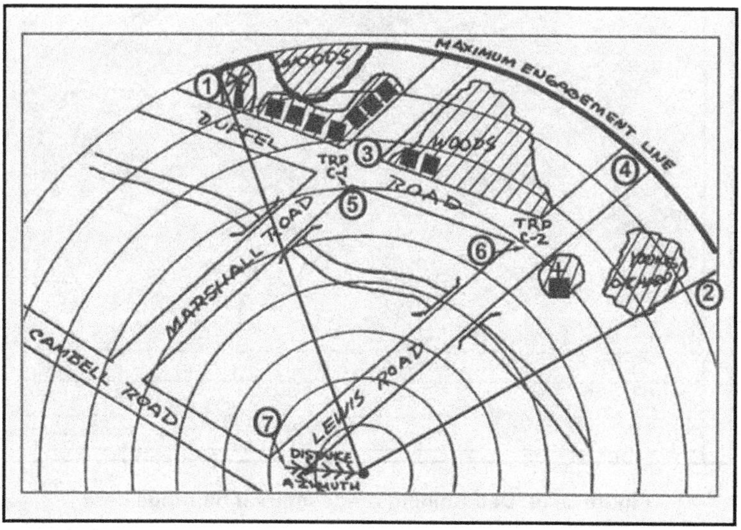

Figure D-8. Placement of distance and azimuth from a known point.

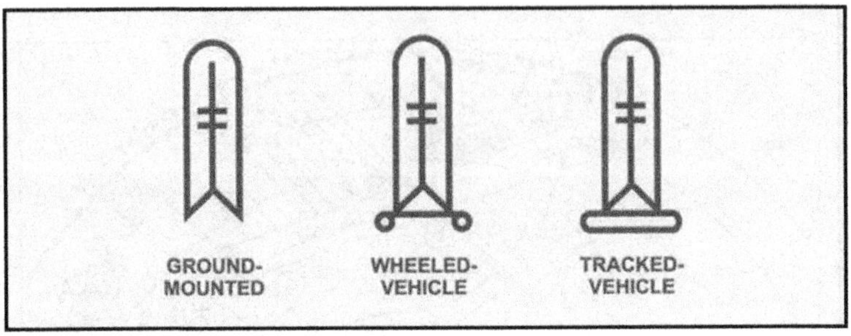

Figure D-9. Weapon symbols.

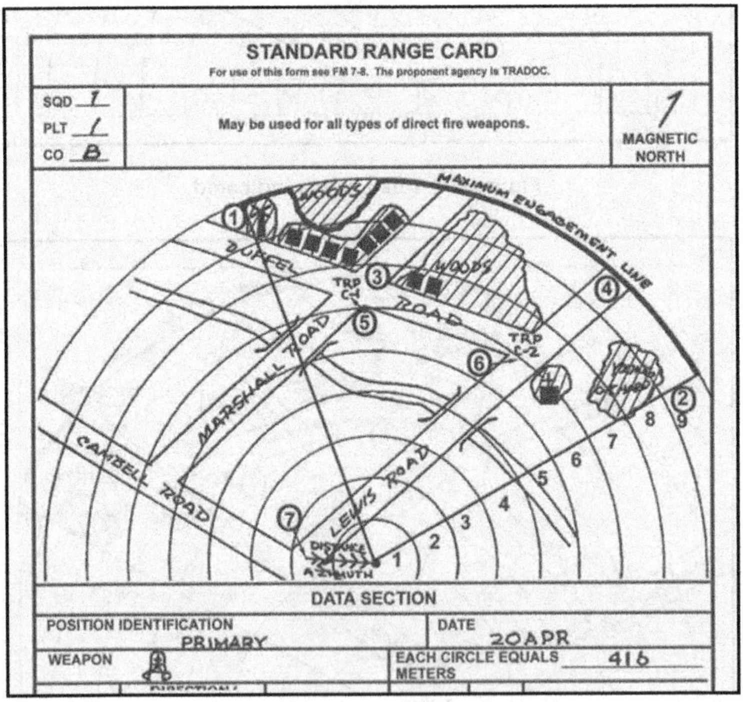

Figure D-10. Determining range interval on range card.

j. Make two range cards. Keep one at the firing position and give one to the squad or section leader for preparation of fire plans and final coordination of fires. (See Figure D-11 for a completed TOW range card.)

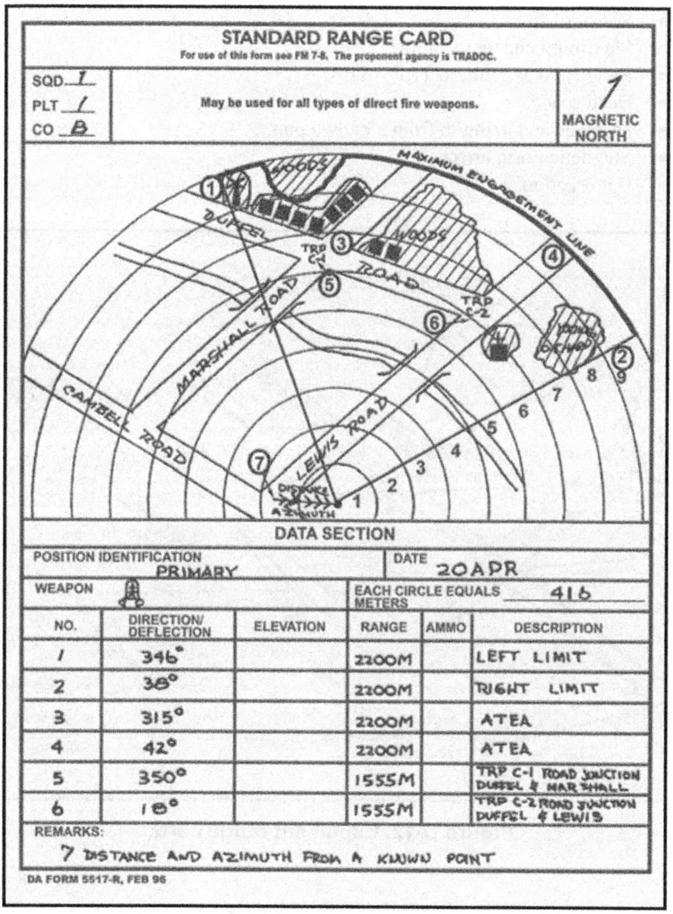
STANDARD RANGE CARD

For use of this form see FM 7-8. The proponent agency is TRADOC.

SQD 1
PLT 1
CO B

May be used for all types of direct fire weapons.

MAGNETIC NORTH

DATA SECTION

POSITION IDENTIFICATION PRIMARY

DATE 20APR

WEAPON

EACH CIRCLE EQUALS 416 METERS

NO.	DIRECTION/ DEFLECTION	ELEVATION	RANGE	AMMO	DESCRIPTION
1	346°		2200M		LEFT LIMIT
2	38°		2200M		RIGHT LIMIT
3	315°		2200M		ATEA
4	42°		2200M		ATEA
5	350°		1555M		TRP C-1 ROAD JUNCTION DUFFEL & MARSHALL
6	18°		1555M		TRP C-2 ROAD JUNCTION DUFFEL & LEWIS

REMARKS:
7 DISTANCE AND AZIMUTH FROM A KNOWN POINT

DA FORM 5517-R, FEB 96

Figure D-11. Example of a completed DA Form 5517-R, Standard Range Card for TOW.

D-3. EXPEDIENT RANGE CARD

In combat, a DA Form 5517-R may not be available. The gunner must then draw a range card on anything available (Figure D-12, page D-10). Preparation of the expedient range card follows the same procedures provided for the standard range card, but the weapon symbol must be used to indicate the location of the weapon position. The range card must include the following eight items:

- Weapon symbol.
- Sector of fire.
- Maximum engagement line.
- Range and azimuth to TRP/ATEA.
- Dead space.
- Distance and azimuth from a known point.
- Magnetic north arrow.
- Data section.

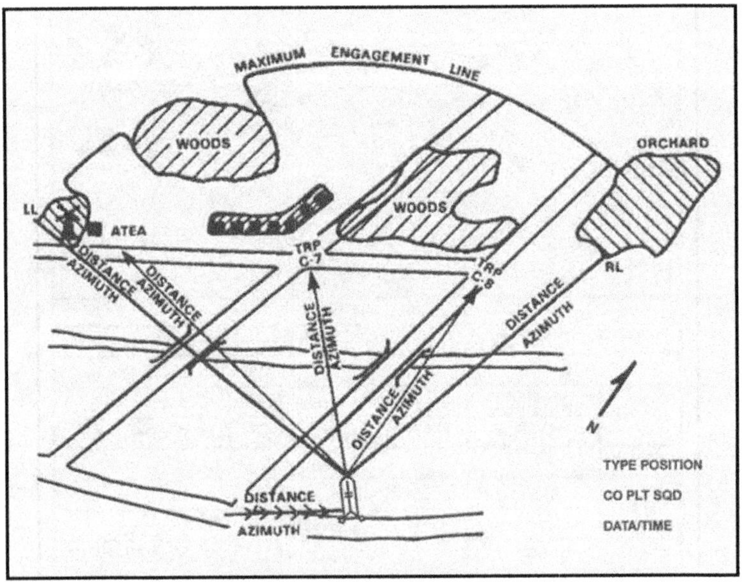

Figure D-12. Expedient range card.

APPENDIX E
SAFETY

Units should develop local directives and SOPs that include individual responsibilities, safety requirements, proximity limits for personnel and explosives, location and sequence of operations, equipment required for handling munitions, and protection for troops. Individual responsibility for operations involving explosives must be designated. Troops must not tamper with the encased missiles other than to remove the forward handling ring and dust cover.

Because of the danger to troops from the backblast, extreme care must be used in all phases of training. Emphasize this danger from the first stage of training. Conduct all crew tasks, position and tracking exercises, and firings with the simulation round as though missiles were being fired.

E-1. SAFETY PRECAUTIONS

The surface danger zone for any firing range consists of a firing area, a target area, impact area, and danger areas surrounding these locations (Figure E-1). An additional area for occupation by personnel during firings may also be required. The shape and size of the surface danger zone varies with the type of missile or rocket being fired. (Refer to DA Pam 385-63 for dimensions.)

a. The primary danger area is a 90-degree cone with a 50-meter radius. The apex of the cone is centered at the rear of the missile launcher. Serious casualties or fatalities are likely to occur to anyone in the area during firing. Hazards include launch motor blast, high noise levels, overpressure, and debris.

b. *Caution area 1* extends radially from each side of the primary danger area to the firing line with a radius of 50 meters. Permanent hearing damage could occur to personnel without adequate hearing protection in this area during firing. The hazards are high noise levels and overpressure.

c. *Caution area 2* is an extension of the primary danger area with the same associated hazards and personnel protection required. The radius of this area is 75 meters.

d. The *200-meter zone* is the danger area for aerial firings 15.25 meters or more above ground level.

e. TOWs will not be fired from buildings or bunkers, or within 100 meters of a vertical or nearly vertical backstop (IAW DA Pam 385-63).

WARNING

All crew members must wear V-51R (or equivalent) earplugs that have been fitted by qualified medical personnel. Failure to use the proper earplugs during missile firing could cause serious injury.

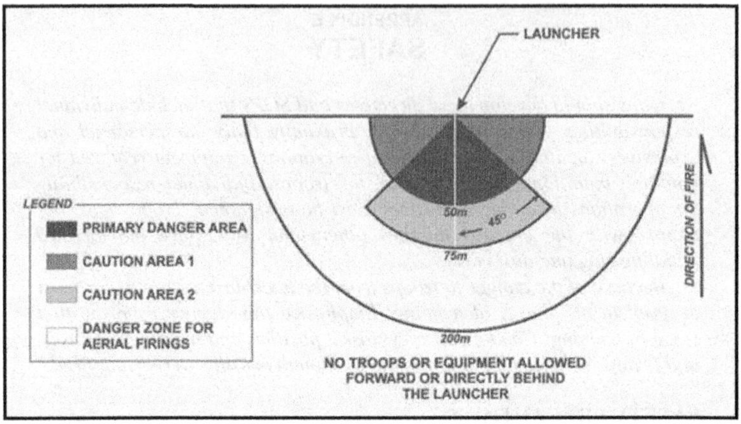

Figure E-1. Surface danger zone for firing basic TOW, TOW 2A, and TOW 2B missiles.

E-2. MISSILE HANDLING PRECAUTIONS

Improper handling of the encased missile may damage the components and cause malfunctions when the missile is launched. If the encased missile is dropped, the end handling rings and the launch container may be damaged. If the missile has been damaged or there is damage to the launch container (other than minor deformation of handling rings), the encased missile should be returned to the ammunition unit for inspection and disposition.

E-3. SIGHTING AND AIMING PRECAUTIONS

Gunners must not look at the sun or bright lights while sighting through the daysight tracker; serious eye burn could result. They must not look through the daysight tracker at an air field test set on the control tower unless the distance between the daysight tracker and the test set is more than 300 feet.

E-4. FIRING ANGLE LIMITATIONS

Azimuth and elevation firing angles are limited by the traversing unit, the vehicle, and other external restrictions. All elevation angles are referenced to the horizontal plane of the traversing unit. All azimuth angles are referenced to the long axis of the vehicle and depend on whether the launch tube points over the front or rear of the vehicle. The other reference line is the line-of-sight from the TOW to the target.

WARNING
At angles greater than 20 degrees above ground level, hazards to the gunner may exist in the overpressure waves and debris caused by the backblast during training. Do not use angles greater than 20 degrees.

a. When the TOW is tripod-mounted, a 360-degree lateral track is possible, because the traversing unit is not restricted in azimuth. Mechanical stops limit the elevation angle coverage to 20 degrees below and 30 degrees above the horizontal plane. Before the missile is fired, the line-of-sight angle should be estimated at the expected time of launch and throughout the expected missile flight time. The firing position should be changed or a different target selected if an expected line-of-sight angle exceeds the firing limitation angle.

b. The firing angle limitations of the TOW carriers are as shown in Figure E-2.

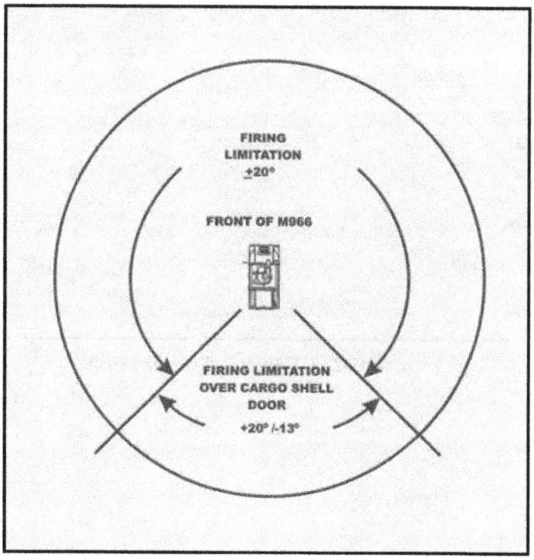

Figure E-2. M966-mounted TOW firing angle limitations.

E-5. SAFETY PRECAUTIONS FOR MILES

The gunner must follow these safety precautions:

a. Do not load an ATWESS cartridge until ready to fire. If the target is lost, remove the ATWESS cartridge from the firing chamber before moving. (Gloves are recommended when loading or unloading the ATWESS cartridge.)

b. Before pulling the PULL-TO-ARM switch, always check to ensure no personnel or equipment (antennas) are in the ATWESS danger zone. This zone extends for 75 meters behind the ATWESS firing chamber and covers an arc 90 degrees wide (Figure E-3).

c. Never view the laser being fired through stabilized optics, such as binoculars or telescopic weapon sights, when within 75 meters of the transmitter.

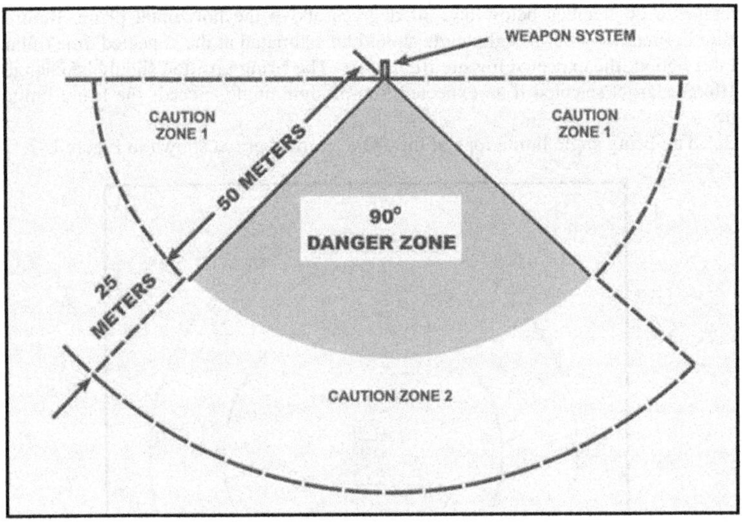

Figure E-3. ATWESS danger zone.

E-6. SAFETY PRECAUTIONS FOR THE TOW GUNNERY TRAINER

The following safety precautions must be observed when using the TGT.

> **DANGER**
>
> **THIS EQUIPMENT USES HIGH VOLTAGE TO OPERATE. NEVER USE UNGROUNDED EXTENSION CORDS, UNGROUNDED ADAPTERS, OR ANY UNGROUNDED OUTLET TO CONNECT THE TGT. DEATH ON CONTACT MAY RESULT IF PERSONNEL FAIL TO OBSERVE SAFETY PRECAUTIONS.**

a. Use two people to lift the instructor console. The console is heavy and lifting with only one man could result in serious injury.

b. Do not attempt to open shipping cases before pressing air pressure release valves on side of cases. Serious injury to personnel could result from opening cases with high pressure inside.

c. Turn off the power to the TGT trainer and disconnect the wall outlet plug before beginning cleaning procedures.

E-7. SAFETY PRECAUTIONS FOR THE TOW FIELD TACTICAL TRAINER

The following safety precautions must be observed when using the TFTT.

a. The laser light emitted by the TFTT is considered eye safe, but suitable precautions must be taken to avoid possible eye damage from overexposure to this radiated energy. The preface to TM 9-6920-453-10 and the laser range safety procedures in DA Pam 385-63 and TB MED-279 discuss these precautions.

b. The M80 blast simulator used with the TFTT can cause death or injury. Observe the precautions listed in the preface of TM 9-6920-453-10.

c. To avoid personnel injury and equipment damage, four people are needed to lift and carry each shipping container.

> **DANGER**
>
> **THIS EQUIPMENT USES HIGH VOLTAGE TO OPERATE. DEATH ON CONTACT MAY RESULT IF PERSONNEL FAIL TO OBSERVE THE SAFETY PRECAUTIONS LISTED IN TM 9-6920-453-10.**

APPENDIX F

TOW TRAINING DEVICES

For many years TOW training devices consisted of the M70-series training sets and MILES TOW equipment. Recently, the Precision Gunnery Training System (PGTS) was fielded. PGTS consists of two systems: an indoor trainer and an outdoor trainer. The TOW Gunner Trainer (TGT) is the indoor system and the TOW Field Tactical Trainer (TFTT) is the outdoor system. The TGT will be used in place of the M70 to qualify individual gunners and for additional sustainment training as needed. The TFTT will be used in place of the M70 for outdoor tracking sustainment and in place of MILES when conducting Tables 5 through 12 of the TOW Training Tables.

MILES TOW equipment is the most realistic device available for simulating tactical engagements. It is valuable in maneuver training exercises and Army training and evaluation programs. However, MILES TOW is not a precision gunnery trainer and should not be used to train gunner tracking skills.

Section I. MULTIPLE-INTEGRATED LASER ENGAGEMENT SYSTEM

MILES is a training system used in force-on-force training and in FTXs. Commanders should use an MPRC to verify MILES equipment before the unit's rotation to the NTC. (See gunnery Tables 7 through 12 in Chapter 4 for target arrays.)

F-1. COMPONENTS AND FEATURES

This paragraph discusses the different components, features, and capabilities of MILES.

a. **Components.**

(1) The following are the nine components in the MILES TOW transit case:

- Operator's manual.
- Two man-worn laser detector sets (halo and harness).
- Two yellow keys.
- TOW laser transmitter/tracker head simulator.
- Missile guidance set adapter.
- Miscellaneous hardware.
- Hook-pile tape pads.
- Control console mounting bracket.
- Collimator adapter bracket.

(2) The following are the ten MILES APC system components packed in the APC transit case.

- Operator's manual.
- Control console.
- Laser transmitter for M2.
- Battery box.
- Hardware.

- Combat vehicle kill indicators.
- Weapon's keys.
- Cable assembly.
- Combat vehicle laser detector belt segments.
- Man-worn laser detectors.

b. **Features.**

Range: 3,750 meters; M60, 1,100 meters.
Tracking time: 10 seconds.

Kill codes:	Visual	Audio
Near miss	2 CVKI flashes	2 tones
Hit	4 to 6 CVKI flashes	4 to 6 tones
Kill	Continuous flash	Continuous tone

Basic load: IAW unit SOP, ATTWESS comes packed 12 to a box.
Backblast area: 75 meters at 90 degrees.
TOW laser tracking: head assembly optics 10 power.
Refire ability: every 10 seconds.

c. **Capabilities.** MILES-equipped weapons have the same range and operational capabilities as the normal weapons. The effective range of the transmitter may be reduced by a dirty transmitter lens and or weak batteries. The M60 machine gun is effective against lightly armored MILES-equipped vehicles and personnel. The TOW 2 is effective against armored MILES-equipped vehicles.

NOTE: Training soldiers to assemble, utilize, and maintain the MILES will greatly decrease down time due to MILES malfunctions.

F-2. SAFETY PRECAUTIONS

The following safety precautions should be observed when using MILES.

a. Do not load an ATWESS cartridge until ready to fire. If the target is lost, remove the ATWESS cartridge from the firing chamber before moving. (Gloves are recommended when loading or unloading the ATWESS.)

b. Before pulling the PULL-TO-ARM switch, always check to ensure no personnel or equipment (antennas) are in the ATWESS danger zone. This zone extends for 75 meters behind the ATWESS firing chamber and covers an arc 90 degrees wide (Figure F-1).

c. Never view the laser being fired through stabilized optics, such as binoculars or telescopic weapon sights, when within 75 meters of the transmitter.

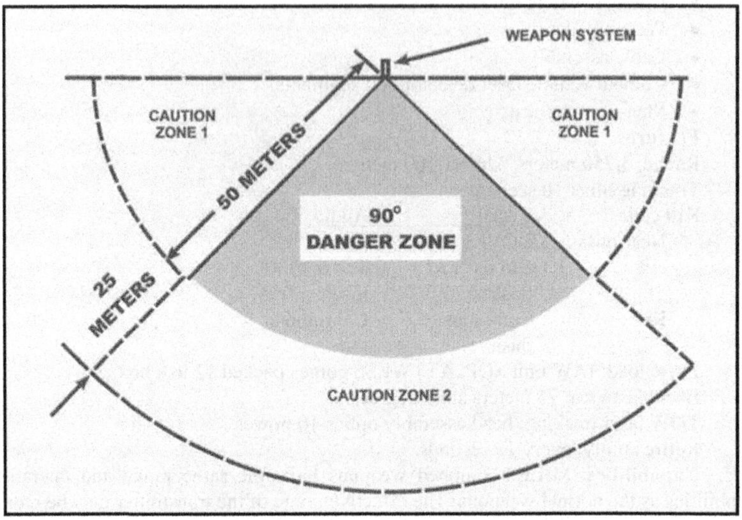

Figure F-1. MILES TOW backblast area.

Section II. TOW GUNNERY TRAINER

The TGT is a new trainer used indoors for individual advanced gunnery tables. It provides realistic gunnery training, not crew drills.

F-3. COMPONENTS AND FEATURES

The TGT consists of TGT-unique components and TOW 2 components. The major TGT-unique components are the instructor station (Figure F-2, page F-4) and the student station (Figure F-3, page F-4). The TOW 2 components are the traversing unit, tripod, and the launch tube.

Figure F-2. Instructor station.

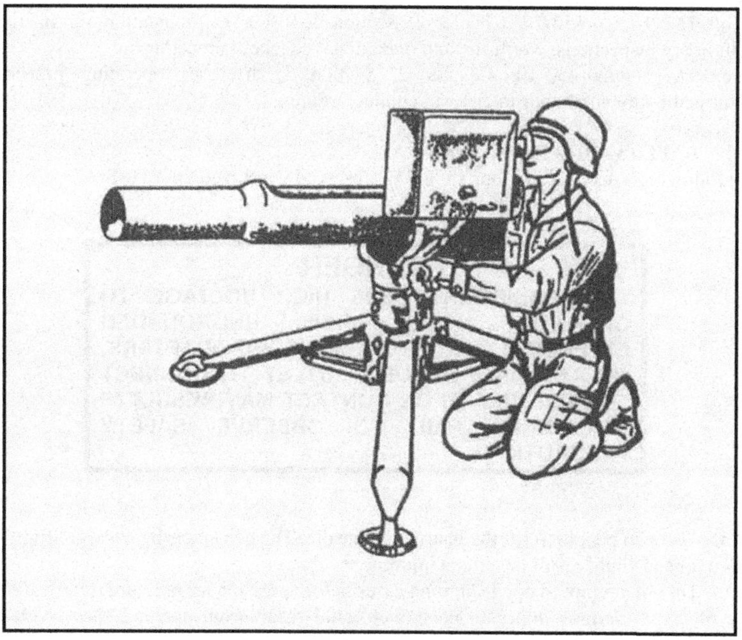

Figure F-3. Student station.

a. The TGT simulates the sight(s), controls, switches and indicators of the M220E4 TOW 2 weapon system. Through the gunner's sight(s) you see battlefield scenes that include both enemy and friendly vehicles. Using the controls and switches, the gunner selects a target, fires, and tracks it. The headsets simulate the blast of the TOW and the singing of the wire. The gunner sees and hears hit-and-miss explosions and hears commands from the instructor and battlefield sounds of small arms and guns. The score is displayed at the end of each mission.

b. TGT trains novice gunnery skill training, gunnery skill progression, and sustainment training. It is used to train the following skills:

- Correct firing position.
- Target identification.
- Target engageability determination.
- Target engagement, including tracking and firing.
- Fire commands.

F-4. ASSEMBLY AND OPERATION

The following manuals are used for proper assembly and operation of the TGT.

a. TM 9-6920-452-10, Chapter 2, Section 1, discusses procedures for unpacking, assembly, disassembly, and repacking for storage and or shipment.

b. TM 9-6920-452-10, Chapter 2, Section 2, discusses preparation for operation, preliminary inspections, warm-up, and operational checkout procedures.

c. TM 9-6920-452-10, Chapter 2, Section 3, discusses operating procedures performed by the instructor in order to conduct training.

F-5. SAFETY PRECAUTIONS

The following safety precautions should be observed when using the TGT.

DANGER

THIS EQUIPMENT USES HIGH VOLTAGE TO OPERATE. NEVER USE UNGROUNDED EXTENSION CORDS, UNGROUNDED ADAPTERS, OR ANY UNGROUNDED OUTLET TO CONNECT THE TGT. DEATH ON CONTACT MAY RESULT IF PERSONNEL FAIL TO OBSERVE SAFETY PRECAUTIONS.

a. Use two people to lift the instructor console. The console is heavy and lifting with only one man could result in serious injury.

b. Do not attempt to open shipping cases before pressing air pressure relief valves on side of cases. Serious injury to personnel could result from opening cases with high pressure inside.

c. Turn off the power to the TGT trainer and disconnect the wall outlet plug before beginning cleaning procedures.

Section III. TOW FIELD TACTICAL TRAINER

The TFTT is used to teach precision gunnery skills to TOW gunners in the field. This training can occur on designated ranges, general outdoor areas, or representative tactical environments. The TFTT trains gunners to adopt a correct firing position, assess target engageability, and to engage and track the target. Missile launch, flight and impact effects are realistically simulated by the TFTT.

F-6. COMPONENTS AND FEATURES

The TFTT consists of TFTT-unique components and a TOW 2 weapon system. The TFTT attaches to the TOW 2 and replaces some of its components. The TFTT can be either tripod mounted or HMMWV mounted and can be configured for the M966 (Figure F-4).

a. The TFTT uses the TOW 2 weapon system equipment to enhance training realism. In addition, most of the TFTT components are designed to resemble actual weapon system equipment. The TFTT uses a retroreflector to designate its target. The retroreflector returns a portion of the laser beam generated by the gunner. The laser beam enables precise measurement of target range and location relative to the gunner. The retroreflector can be mounted on a variety of target vehicles, which can be maneuvered as required during a training mission. Targets equipped with MILES sensors can also be engaged by the TFTT.

b. The TFTT operator loads the M80 blast simulator, sets the duration of the obscuration that simulates the smoke produced at missile launch, and selects the relative size of the target. Following missile launch, the operator monitors gunner performance during missile flight. At the end of each mission, the operator is provided with a readout of mission results.

F-7. ASSEMBLY AND OPERATION

The following manuals are used for proper assembly and operation of the TFTT.

a. TM 9-6920-453-10 (M966), Chapter 2, Section 1, discusses procedures for unpacking, assembly, disassembly, and repacking of the TFTT.

b. TM 9-6920-453-10 (M966), Chapter 2, Section 2, discusses procedures for preliminary inspection, power-up, and operational checkout of the TFTT.

c. TM 9-6920-453-10 (M966), Chapter 2, Section 3, discuss procedures for operating the TFTT.

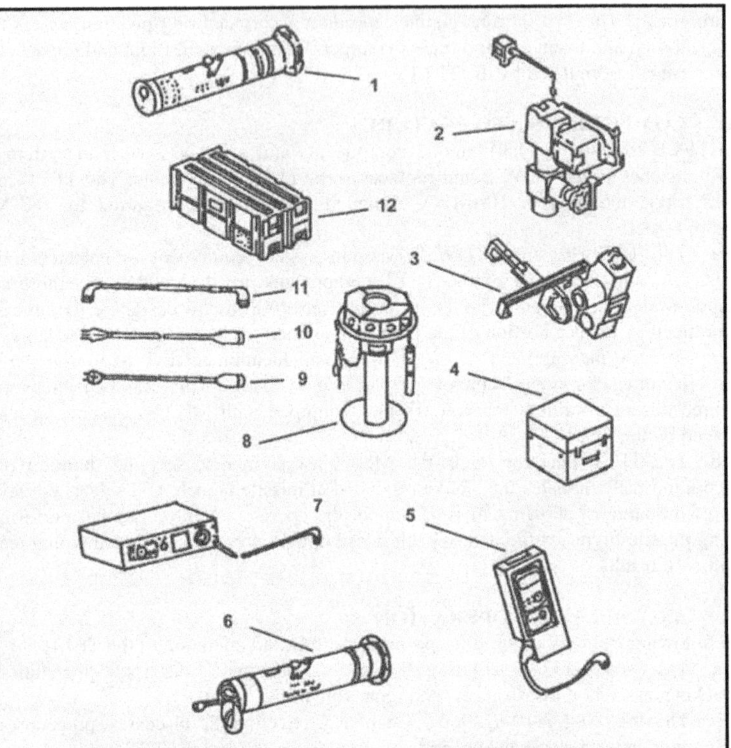

TRAINER MISSILE TUBE (1)
DAYSIGHT TRACKER INTERFACE (2)
THERMAL SIGHT INTERFACE (3)
COMPUTER (4)
REMOTE CONTROL UNIT (5)
BATTERY TUBE (6)
CONVERTER/CHARGER (7)
RETROFLECTOR ASSEMBLY AND PEDESTAL (8)
CABLES (9, 10, 11)
SHIPPING CONTAINERS (12)

Figure F-4. TOW field tactical trainer components (M966).

F-8. SAFETY PRECAUTIONS

The laser light emitted by the TFTT is considered eye safe, but suitable precautions must be taken to avoid possible eye damage from overexposure to this radiated energy. See the preface to TM 9-6920-453-10 and the laser range safety procedures in DA Pam 385-63 and TB MED-279 for these precautions. To avoid personnel injury and equipment damage, four people are required to lift and carry each shipping container.

DANGER

THIS EQUIPMENT USES HIGH VOLTAGE TO OPERATE. DEATH ON CONTACT MAY RESULT IF PERSONNEL FAIL TO OBSERVE SAFETY PRECAUTIONS LISTED IN TMs 9-6920-453-10 (M966).

DANGER

THE M80 BLAST SIMULATOR USED WITH THE TFTT CAN CAUSE DEATH OR INJURY. OBSERVE THE PRECAUTIONS LISTED IN THE PREFACE OF TM 9-6920-453-10.

APPENDIX G

COMBAT VEHICLE IDENTIFICATION

Skill Level 1 11B soldiers must be able to identify combat vehicles according to the conditions and standards listed in this appendix.

CONDITIONS: Given a classroom with a 35-mm slide projector, screen, slide tray with 40 combat vehicle slides, stopwatch, chair, desk or clipboard, and pencil. (See the local Training Support Center for the combat vehicle slides.)

STANDARDS: Given a 12-second exposure time for each slide, the soldier will identify 36 of 40 vehicles by nomenclature.

EVALUATION PROCEDURES: This task requires the crewman to identify combat vehicles from the front, side, oblique, or rear using day and thermal sights. Views of these combat vehicles should replicate ranges between 800 and 1,200 meters.

a. **Administrative Process.** At the station, the evaluator logs the soldier's information on a roster. The evaluator provides the soldier with all materials and equipment and displays them in a manner outlined in this guide. The evaluator must use the criterion performance checklist for this task. The evaluator informs the soldier of his performance on this task and directs him to the next station or to further training. The number of soldiers to be tested at one time depends on the classroom seating without compromising test procedures.

b. **Administrative Procedures for Personnel Receiving a NO-GO.** If the soldier does not achieve the standard indicated on the checklist, he receives a NO-GO. When a soldier receives a NO-GO, he must be critiqued, to include an explanation of vehicles that were incorrectly identified. The soldier will be retested with a second set of slides showing the same vehicles in a different view or a different order.

PERSONNEL, EQUIPMENT, AND MATERIAL REQUIRED:

- Qualified 11B sergeant or above.
- Forty 35-mm slides of combat vehicles. The 30 required vehicles for this test are included in the list in Figure H-1 and are indicated by an asterisk. The balance will be taken from the other vehicles listed in Figure G-1 (page G-4). (Recognition of Combat Vehicles [ROC-V] compact discs may also be used.) (For more information, see GTA 17-02-013.)
- Classroom.
- Slide projector with screen.
- Stopwatch.
- Performance checklist (one per soldier).
- Chair, desk or clipboard (one per soldier).
- Pencil (one per soldier).

PRETEST PREPARATION: Thirty vehicles will be taken from the prescribed list and ten will be selected by the commander based on the unit's contingency area. The station evaluator must ensure the slide projector is operational and the slides can be recognized from all locations in the classroom. Some projectors have an automatic setting for exposure of each slide; if used, this feature must be checked for accuracy.

a. Select 30 required vehicle slides to be used along with 10 slides selected by the commander.

b. Ensure the slide projector is operational.

c. Ensure the slides can be recognized from all locations in the classroom.

PRETEST CONDITIONS FOR EACH EXAMINEE: The slide projector will be on, but slides will not be exposed.

TEST PLANNING TIME:

Administrative: 5 minutes
Test: 15 minutes
Total: 20 minutes

INSTRUCTIONS TO EXAMINEES:

"LET ME HAVE YOUR ATTENTION. AT THIS STATION, YOU WILL BE TESTED ON YOUR ABILITY TO IDENTIFY COMBAT VEHICLES. YOU MUST CORRECTLY IDENTIFY THIRTY-SIX OF FORTY VEHICLES BY NOMENCLATURE. YOU WILL HAVE TWELVE SECONDS TO VIEW EACH SLIDE AND TEN SECONDS BETWEEN EACH SLIDE TO WRITE YOUR ANSWER ON THE PERFORMANCE CHECKLIST PROVIDED. BE SURE YOU DO NOT GET OUT OF SEQUENCE OR YOUR ANSWERS WILL BE INCORRECT. DO YOU UNDERSTAND THESE INSTRUCTIONS?"

Pause 5 seconds, then say, "YOU WILL HAVE TWENTY-TWO SECONDS TO COMPLETE EACH TASK." Pause, then say, "BEGIN."

Begin timing when the command "BEGIN" is given. If the examinee has not completed each task after the appropriate time, announce loud enough for the examinee to hear, "STOP."

PERFORMANCE CHECKLIST

TASK: Identify combat vehicles by nomenclature.

PERFORMANCE MEASURES:

1. Identified vehicle slide No. 1.
2. Identified vehicle slide No. 2.
3. Identified vehicle slide No. 3.
4. Identified vehicle slide No. 4.
5. Identified vehicle slide No. 5.
6. Identified vehicle slide No. 6.
7. Identified vehicle slide No. 7.

8. Identified vehicle slide No. 8.
9. Identified vehicle slide No. 9.
10. Identified vehicle slide No. 10.
11. Identified vehicle slide No. 11.
12. Identified vehicle slide No. 12.
13. Identified vehicle slide No. 13.
14. Identified vehicle slide No. 14.
15. Identified vehicle slide No. 15.
16. Identified vehicle slide No. 16.
17. Identified vehicle slide No. 17.
18. Identified vehicle slide No. 18.
19. Identified vehicle slide No. 19.
20. Identified vehicle slide No. 20.
21. Identified vehicle slide No. 21.
22. Identified vehicle slide No. 22.
23. Identified vehicle slide No. 23.
24. Identified vehicle slide No. 24.
25. Identified vehicle slide No. 25.
26. Identified vehicle slide No. 26.
27. Identified vehicle slide No. 27.
28. Identified vehicle slide No. 28.
29. Identified vehicle slide No. 29.
30. Identified vehicle slide No. 30.
31. Identified vehicle slide No. 31.
32. Identified vehicle slide No. 32.
33. Identified vehicle slide No. 33.
34. Identified vehicle slide No. 34.
35. Identified vehicle slide No. 35.
36. Identified vehicle slide No. 36.
37. Identified vehicle slide No. 37.
38. Identified vehicle slide No. 38.
39. Identified vehicle slide No. 39.
40. Identified vehicle slide No. 40.

TANKS

1. AMX13
2. AMX30
3. AMX30 SA
4. JP Kanone
5. Centurion
6. Centurion 15
7. Centurion RE
*8. Challenger
*9. Chieftain
10. Leopard 1
*11. Leopard 2
*12. M1 Abrams
13. M48A5
14. M551 Sheridan
15. M60A1
*16. M60A3
*17. Merkava
18. PT76
19. PT76 Type 60
20. T8OU
21. T34 Medium
22. T54
23. T55
24. T55K
*25. T62
26. T64
27. T64A
28. T64B
29. T64K
*30. T72
31. T72 M84
32. T72M1
*33. T80
34. Type 59
35. Type 62
36. Type 69
37. Type 79
38. Type 80
39. Type 90
*40. Le Clerc

LIGHT-ARMOR—APC/IFV/RECONNAISSANCE

1. AML
2. AMX10
3. AMX10P
4. AMX10RC
5. BMD1
6. BMD 1979/3
*7. BMD2
*8. BMP1
9. BMP1K
10. BMP1KSH
*11. BMP2
*12. Bradley M2/M3
13. AMX VCI
14. Bradley M2A2
15. LVTP7 (AAV7A1)
16. Bradley M3A1
17. Bradley M3A2
*18. BRDM1
*19. BRDM2
20. BRDM2 AT3
21. BRDM2 AT5
22. BRDM2 RKH
23. BRDM2 SA9
24. BRDM2 U
25. BRM
26. BRM1
27. BTR152
28. KM900
29. BTR50PK
30. BTR60PA Cmd
*31. BTR60PB
*32. BTR70
33. BTR80
34. BTRD
35. Charrua
36. Condor
37. Cougar
38. EE-9 Cascavel
39. EE-11
40. Fox
41. FV432
42. IPR
43. IRM
44. Jaguar 1
45. Jaguar 2
*46. Luchs
*47. M113
48. M113A3
49. M3 APC (Fr)
*50. Marder
51. MCT (S)
52. Ferret MK 1
53. Ferret MK 2
54. MTLB
55. NFV-1
56. NVH-1
57. OT64
58. OT65
59. PSZH-IV
60. Ratel
61. Rooikat
62. Saxon
*63. Scimitar
*64. Scorpion
65. Striker
*66. VAB
*67. Warrior
68. WZ523
69. WZ551
70. YW531
71. YW534

*Required for test.

Figure G-1. Combat vehicle slide list.

ANTITANK SYSTEMS

1. ASU57
2. ASU85
3. AT1 Snapper
4. AT2 Swatter
5. AT3 Sagger
6. AT4 Spigot
7. AT5 Spandrel
8. AT6 Spiral
9. B10
10. PANHARD VCR HOT
*11. ITV M901
12. Milan ATGM
13. MT12
14. RPG2
15. RPG7
16. SPG9
17. SS11 Harpoon ATGM
18. Swingfire ATGM
19. Type 52 75-mm Recoilless Rifle
20. Type 56 RPG-2
21. Type 70-1 62-mm Rocket Launcher
22. Type 86

ANTIAIRCRAFT

1. AMX DCA 30-2
2. Chaparral
*3. Gepard
4. M163A1 VULCAN
5. Roland
6. S60
7. ZPU4
8. 2S6
*9. ZSU23-4
10. ZSU57-2
11. ZU23

SELF-PROPELLED MORTAR

1. M106 107-mm
2. M125 81-mm
3. T54 160-mm
4. Type 85 120-mm
5. Type 85 82-mm
6. YW304 82-mm
7. YW381 120-mm

*Required for test.

Figure G-1. Combat vehicle slide list (continued).

SELF-PROPELLED AND TOWED ARTILLERY

1. 2S1
2. 2S3
3. 2S4
4. 2S5
5. 2S7
6. 2S9
7. AL FAO 210-mm
8. Astros (MRL) 127-mm
9. Astros (MRL) 180-mm
10. Astros (MRL) 300-mm
11. BM21
12. CGT F1
13. Dana 152-mm (SP)
14. G-5 155-mm towed
15. G-6 155-mm (SP)
16. GCT 155-mm (SP)
17. GHN-45 155-mm towed
18. M107
19. M109A1
20. M109A2
21. M109A3
22. M109A6
23. M110A2
24. M1972 122-mm
25. M1973 152-mm
26. M1974 152-mm (SP)
27. M1975 130-mm (SP)
28. M1977 122-mm (SP)
29. M1978 170-mm (SP)
30. M1981 122-mm (SP)
31. M1985 MRL 122-mm (SP)
32. M1985 MRL 240-mm (SP)
33. M1989 170-mm
34. Majnoon 155-mm (SP)
35. MKF3
36. MLRS
37. Type 54-1 122-mm
38. Type 59 130-mm
39. Type 85 122-mm

MISCELLANEOUS

1. Artillery Command and Reconnaissance Vehicle
2. AVLB
3. Fox, NBC Reconnaissance M93
* 4. LAV-25
5. M728 CEV
* 6. M88A1
7. M9 ACE
8. PRP-3

*Required for test.

Figure G-1. Combat vehicle slide list (continued).

APPENDIX H

OPFOR/THREAT COUNTERPART SYSTEMS

Recent years have seen rapid and revolutionary changes in the field of armor and antiarmor warfare. The most radical changes have been in the design of armor itself, the fire control systems (especially in thermal sights), the warheads and kill mechanisms of the missiles, and the various countermeasures that have been developed. The rate of change is increasing, and antiarmor crewmen and trainers must make a continual effort to keep abreast of recent developments. What was true two years ago, or even one year ago, is not necessarily true today. Leaders who take their men into battle and make decisions based on outmoded information and assumptions are dooming those men to failure and death.

H-1. OPFOR/THREAT ARMORED VEHICLES

Armored vehicle identification in the past was a simple matter of distinguishing between vehicles manufactured by Eastern Bloc and Rest of the World (ROW) nations and those manufactured by NATO nations. Even then, many soldiers had difficulty distinguishing between vehicles on a simple "friend or foe" basis. Changes in international relations have made it less likely that we will meet nations of the former Soviet Union in battle, but more likely that we will fight the military of a third world nation equipped with a wide variety of equipment. From the standpoint of vehicle identification, this greatly complicates the problem. Most third world nations do not manufacture their own vehicles, but buy them from other countries. As a result, many of them have vehicles from nearly every manufacturing nation in the world in their inventories. Twenty nations in the world today have more than 500 main battle tanks each. Two possible examples of many illustrate this point (Table H-1).

COUNTRY	TYPE	ORIGIN
Iran	Type 59 Type 69 Chieftain M48A5 M60A1 T62 T72 EE-9 Cascavel M113	China China United Kingdom USA USA Russia Russia Brazil USA
Iraq	Type 59 AMX-30 Chieftain T62 T72 EE-11 Urutu Walid MOWAG Roland M113A1	China France United Kingdom Russia Russia Brazil Egypt Switzerland USA

Table H-1. Select armored vehicle types and countries of origin.

a. The task of armored vehicle identification has become much more complex. The antiarmor crewman must now become familiar with a much wider variety of vehicles than before. He must also be well informed about what vehicles a potential enemy possesses. In the future, the antiarmor crewman will probably face armored vehicles produced in the U.S. as well as in other friendly nations. The trend of the U.S. to fight with a coalition means that, on future battlefields, both friendly and enemy forces may have identical equipment.

b. This manual is not a complete FM on vehicle identification. However, the antiarmor crewman should be readily familiar with a few of the most recently fielded vehicles and the particular problems and capabilities they present.

(1) ***British Main Battle Tank, Chieftain Mk 5*** (Figure H-1).

- Date of introduction: 1967 original Chieftain
- Proliferation: At least six countries
- Automotive performance:
  Engine type: 750-hp diesel
  Cruising range: 400 to 500 km
- Speed:
  Max road: 48 km/h
  Max off-road: INA
  Average cross country: 30 km/h
  Max swim: INA
- Fording depths: 1.1 meters unprepared
- Main armament ammunition:
  Caliber type name: 120-mm high L11A5
  Rate of fire: 8 to 10 rounds per minute first minute; 6 sustained
  Maximum aimed range: 5,000 meters
  Maximum effective range:
    Day: 3,000 meters
    Night: INA
  Armor penetration: INA

Figure H-1. British main battle tank, Chieftain Mk 5.

(2) ***German Main Battle Tank, Leopard 2*** (Figure H-2).

- Date of introduction: 1979
- Proliferation: At least seven countries
- Automotive performance:
  Engine type: 750-hp diesel
  Cruising range: 400 to 500 km
- Speed:
  Max road: 48 km/h
  Max off-road: INA
  Average cross country: 30 km/h
  Max swim: INA
- Fording depths: 1.1 meters unprepared
- Main armaments ammunition:
  Caliber type, name: 120-mm APFSDS-T, DM43
  Rate of fire: INA
  Maximum aimed range: 3,500 meters
    Day: INA
    Night: INA
  Armor penetration: 450 to 2,000 mm

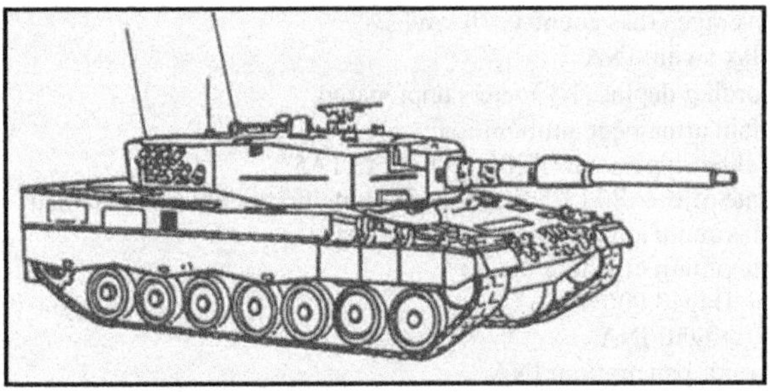

Figure H-2. German main battle tank, Leopard 2.

(3) ***Polish/Czechoslovakian Main Battle Tank, T-72M1*** (Figure H-3, page H-4).

- Date of introduction: 1975
- Proliferation: At least seven countries
- Automotive performance:
  Engine type: 780-hp diesel
  Cruising range: 460 to 700 km with extra tanks
- Speed:
  Max road: 60 km/h
  Max off-road: 45 km/h
  Average cross country: 35 km/h
  Max swim: INA
- Fording depths: 1.2 meters unprepared; 5 meters with snorkel

- Main armaments ammunition:
  Caliber type, name: 125-mm APFSDS-T, BM-42M
  Rate of fire: INA
  Maximum aimed range: 2,000 to 3,000 meters
    Day: 2,000 to 3,000 meters
    Night: 850 to 1,300 meters
  Armor penetration: 590 to 630 rounds per min at 2,000 meters

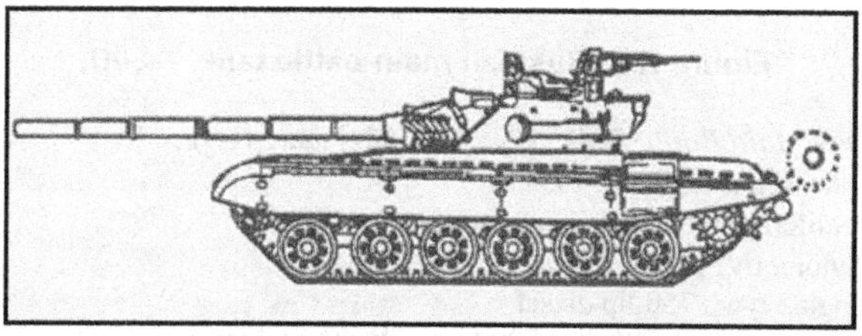

Figure H-3. Polish/Czechoslovakian main battle tank, T-72M1.

(4) ***Russian Main Battle Tank, T-80U*** (Figure H-4).

- Date of introduction: 1987
- Proliferation: At least three countries
- Automotive performance:
  Engine type: 1250-hp gas turbine (multi-fuel); diesel on T-80UD
  Cruising range: 335 km; 600 km with extra tanks
- Speed:
  Max road: 70 km/h
  Max off-road: 48 km/h
  Average cross country: 40 km/h
  Max swim: INA
- Fording depths: 1.8 meters unprepared; 5 meters with snorkel; 12 meters with BROD-M system
- Main armaments ammunition:
  Caliber type, name: 125-mm APFSDS-T, BM-42M
  Rate of fire: INA
  Maximum aimed range: 2,000 to 3,000 meters
    Day: 2,000 to 3,000 meters
    Night: 800 to 1,300 meters
  Armor penetration: 560 to 630 mm at 2,000 meters

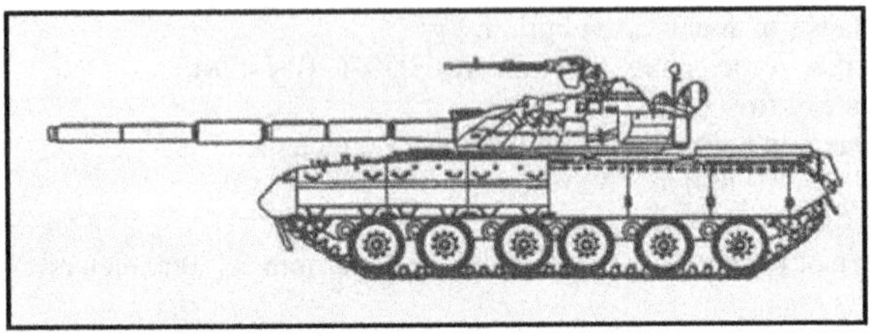

Figure H-4. Russian main battle tank, T-80U.

(5) ***Chinese Main Battle Tank, Type 85-IIM*** (Figure H-5).

- Date of introduction: 1991
- Proliferation: At least two countries
- Automotive performance:
  Engine type: 750-hp diesel
  Cruising range: 700 km; 900 km with external tanks
- Speed:
  Max road: 57 km/h
  Max off-road: 45 km/h
  Average cross country: 35 km/h
  Max swim: INA
- Fording depths: 1.4 meters unprepared; 2.4 meters with snorkel
- Main armaments ammunition:
  Caliber type, name: 125-mm HEAT-MP, BK-29M
  Rate of fire: INA
  Maximum aimed range: 3,000 meters
    Day: INA
    Night: 850 to 1,300 meters
  Armor penetration: 560 to 750 mm

Figure H-5. Chinese main battle tank, Type 85-IIM.

H-2. THREAT RIFLE-PROPELLED GRENADES

The most prolific threat now is the hand-held RPG. The following are some of the RPGs that U.S. forces will encounter.

a. **Russian 40-mm Antitank Grenade Launcher, RPG-7V** (Figure H-6).

- Date of introduction: 1962
- Proliferation: At least 40 countries
- Ammunition: PG-7V
  Caliber: 85 mm
  Type: Heat
  Range:
    Effective: 500 meters
    Minimum: N/A
    Moving: 300 meters
  Armor penetration: 330 mm

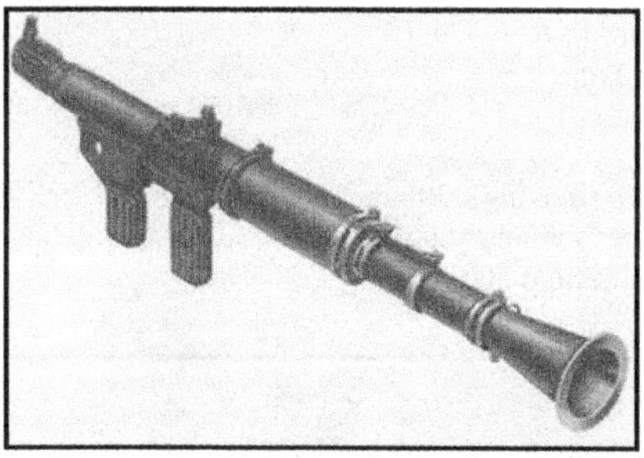

Figure H-6. Russian 40-mm antitank grenade launcher, RPG-7V.

b. **Swedish 84-mm Recoilless Rifle, Carl Gustav M2** (Figure H-7).

- Date of introduction: INA
- Proliferation: At least 20 countries
- Ammunition: FFV 751
  Type: Heat (tandem)
  Range:
    Effective: 500 meters
    Minimum: N/A
  Armor penetration: +500 mm

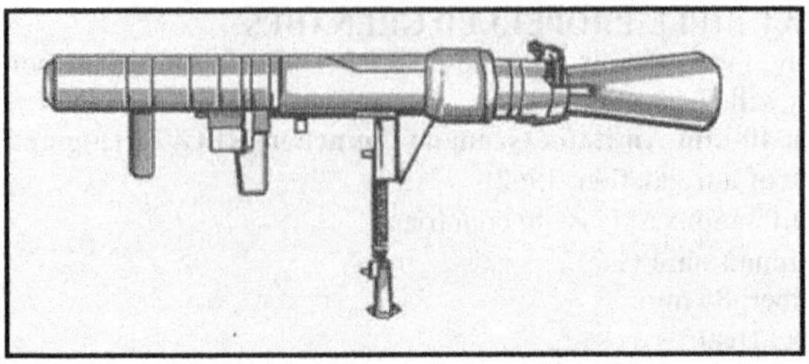

Figure H-7. Swedish 84-mm recoilless rifle, Carl Gustav M2.

c. **German 60-mm Antitank Grenade Launcher, Panzerfaust-3** (Figure H-8).

- Date of introduction: 1990
- Proliferation: At least eight countries
- Ammunition: INA
  Caliber: 110 mm
  Type: Heat
  Range:
    Effective (moving): 300 meters
    Effective (stationary): 500 meters
  Armor penetration: 500 mm

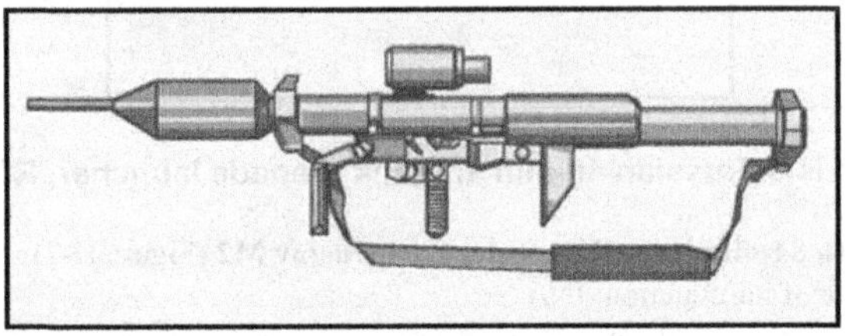

Figure H-8. German 60-mm antitank grenade launcher, Panzerfaust-3.

d. **German 67-mm Disposable Antitank Grenade Launcher, Armbrust** (Figure H-9, page H-8).

- Date of introduction: INA
- Proliferation: At least seven countries
- Ammunition: INA
  Type: Heat
  Range:
    Maximum: 1500 meters
    Effective AT: 300 meters

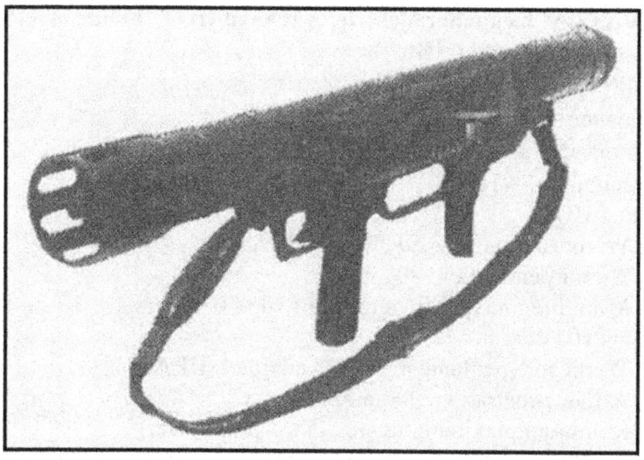

Figure H-9. German 67-mm disposable antitank grenade launcher, Armbrust.

e. **Russian 105-mm Antitank Grenade Launcher, RPG-29** (Figure H-10).

- Date of introduction: Late 1980s
- Proliferation: Former Soviet Union
- Ammunition: PG-29V

  Caliber (warhead): 105 mm

  Type: Heat (tandem)

  Range:

    Effective: 500 meters

    Minimum: N/A

  Armor penetration: +750 mm

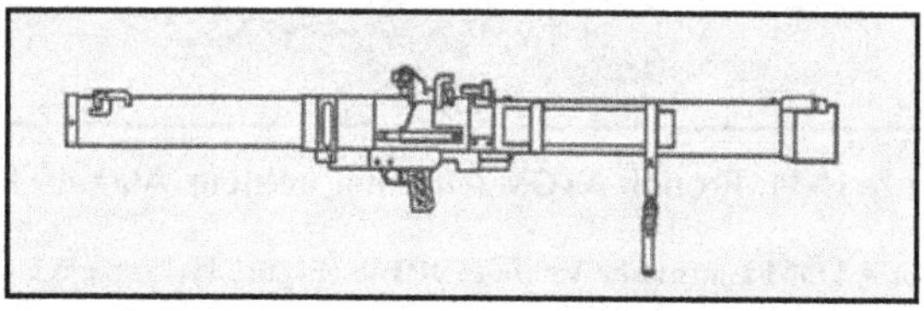

Figure H-10. Russian 105-mm antitank grenade launcher, RPG-29.

H-3. ANTITANK GUIDED MISSILE LAUNCHER VEHICLES

Although there are unique ATGM launcher vehicles with unique ATGMs, most launcher vehicles are military and commercial vehicles adapted with pintel mounts for portable ground launchers, with ATGMs manually loaded and launched. The systems described in this paragraph are the more common threat systems or represent the spectrum of antitank systems that can threaten U.S, Army forces in the world today.

a. **French ATGM Launcher Vehicle, AMX-10 HOT** (Figure H-11).

- Date of introduction: INA
- Proliferation: At least one country
- Guidance system: SALCOS
- Command link: Wire
- Ammunition: ATGM
  Name: HOT
    Warhead type: shaped charge (HEAT)
    Armor penetration: 800 mm
    Minimum/maximum range: 75 to 4,000 meters
  Name: HOT 2
    Warhead type: tandem shaped charged (HEAT)
    Armor penetration: 900 mm
    Minimum/maximum range: 75 to 4,000 meters
  Name HOT 2
    Warhead type: tandem shaped charged (HEAT)
    Armor penetration: 1,250 mm
    Minimum/maximum range: 75 to 4,000 meters
- Probability of hit: INA
- Time of flight to max range: 17.3 seconds

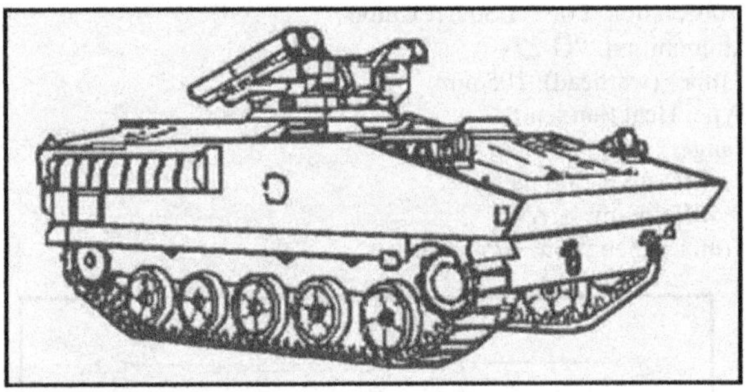

Figure H-11. French ATGM launcher vehicle, AMX-10 HOT.

b. **Russian ATGM Launcher Vehicle, 9P148** (Figure H-12, page H-10).

- Date of introduction: 1977
- Proliferation: At least six countries
- Guidance system: SALCOS
- Command link: Wire
- Ammunition
- ATGM:
  Name: AT-5/SPANDREL
  Warhead type: shaped charge (HEAT)
  Armor penetration: 650 mm
  Minimum/maximum range: 75 to 4,000 meters

- Probability of hit: 90 percent
- Time of flight to max range: 20 seconds

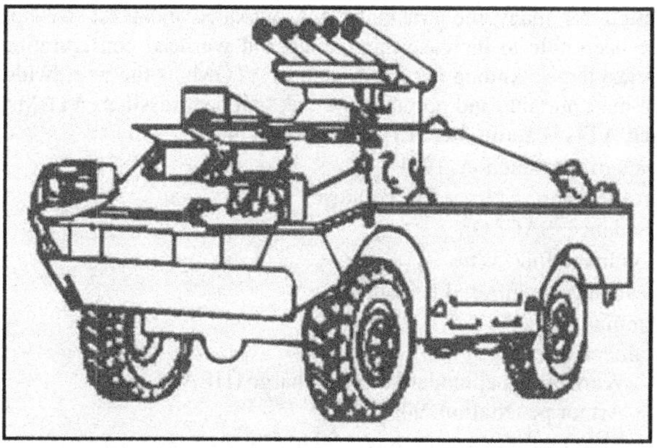

Figure H-12. Russian ATGM launcher vehicle, 9P148.

c. **Russian ATGM Launcher Vehicle, 9P149** (Figure H-13).

- Date of introduction: 1990
- Proliferation: At least nine countries
- Guidance system: SALCOS
- Command link: radio frequency
- Ammunition: ATGM
  Name: AT-56a/SPIRAL
    Warhead type: shaped charge (HEAT)
    Armor penetration: 750 mm
    Minimum/maximum range: 400 to 5,000 meters
- Probability of hit: 90 percent
- Time of flight to max range: 14.5 seconds

Figure H-13. Russian ATGM launcher vehicle, 9P149.

H-4. ANTITANK GUIDED MISSILE LAUNCHER

As the levels of armor protection and the lethality of antitank weapons continue to rise, armor protection for many modern tanks has outpaced most AT weapons. However, ATGMs have been able to increase their range and warhead configurations to threaten even the heaviest tanks. Among notable trends in ATGMs is the worldwide proliferation and variety of man-portable and portable antitank ballistic missiles (ATBMs).

a. **French ATGM Launcher, Eryx** (Figure H-14).

- Date of introduction: 1991
- Proliferation: At least five countries
- Guidance: SACLOS
- Command link: Wire
- Beacon type: infrared laser diode
- Ammunition:
  Name: Eryx
    Warhead type: tandem shaped charge (HEAT)
    Armor penetration: 900 mm
    Minimum/maximum range: 50 to 600 meters
- Probability of hit: 90 percent
- Time of flight to max range: 3.7 seconds

Figure H-14. French ATGM launcher, Eryx.

b. **Russian ATGM Launcher, AT-4/AT-5** (Figure H-15, page H-12).

- Date of introduction: 1973
- Proliferation: At least 25 countries
- Guidance: SACLOS
- Command link: Wire
- Beacon type: Incandescent infrared bulb
- Ammunition:

Name: AT-5B/SPANDREL-B
    Warhead type: tandem shaped charge (HEAT)
    Armor penetration: 925 mm
    Minimum/maximum range: 75 to 4,000 meters
    Probability of hit: 90 percent
    Time of flight to max range: 10 seconds
Name: AT-5B/SPANDREL
    Warhead type: shaped charge (HEAT)
    Armor penetration: 650 mm
    Minimum/maximum range: 75 to 4,000 meters
    Probability of hit: 90 percent
    Time of flight to max range: 20 seconds
Name: AT-4/SPIGOT
    Warhead type: shaped charge (HEAT)
    Armor penetration: 480 mm
    Minimum/maximum range: 70 to 4,200 meters
    Time of flight to max range: 11 seconds

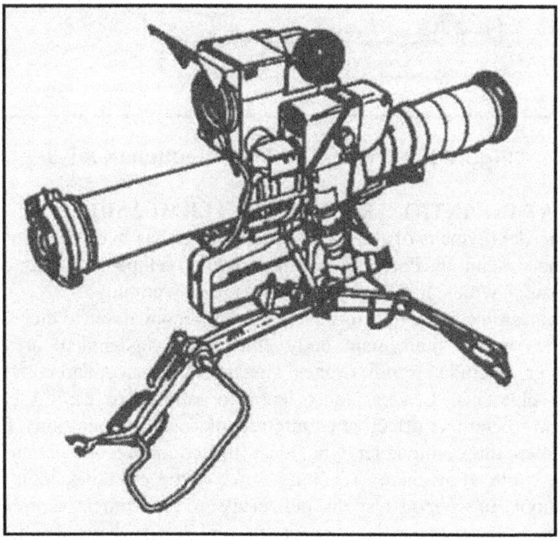

Figure H-15. Russian ATGM launcher, AT-4/AT-5.

c. **Russian ATGM Launcher, AT-3** (Figure H-16).

- Date of introduction: 1963
- Proliferation: At least 45 countries
- Guidance: MCLOS
- Command link: Wire

- Beacon type: Incandescent infrared bulb (SACLOS)
- Ammunition:
  Name: AT-3, 3a, 3b/SAGGER
    Warhead type: tandem shaped charge (HEAT)
    Armor penetration: 400 mm
    Minimum/maximum range: 500 to 3,000 meters
    Probability of hit: 70 percent
    Time of flight to max range: 26 seconds

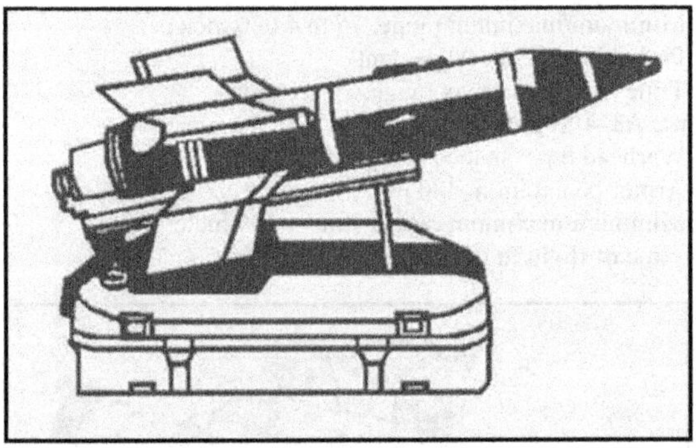

Figure H-16. Russian ATGM launcher, AT-3.

H-5. THREAT INFANTRY ARMOR COUNTERMEASURES

Historically, the deployment of thick belts of ATGMs has been the foundation of U.S. antiarmor defense planning. Potential enemies have developed countermeasures to these antiarmor measures, which any enemy we may face may employ.

a. **Reconnaissance.** A high priority for enemy reconnaissance elements is to locate our ATGM sites before their main body enters our engagement areas. Methods of reconnaissance can include aircraft, drones, signals intelligence, and conventional ground reconnaissance elements. Leaders must learn to emphasize careful camouflage and concealment and to conduct effective counterreconnaissance operations. Lessons learned at NTC show that the commander who loses the counterreconnaissance phase of the battle loses the battle. For enemy reconnaissance to be effective, their reconnaissance elements must spot our weapon systems and survive to report their location.

b. **Artillery.** Once enemy reconnaissance elements have spotted our ATGM positions, the weapon of choice to destroy them is artillery. Since ATGMs are a high-priority target for artillery, these positions will be subjected to intense indirect fire. The TOW is not designed to withstand such barrages. The best way to survive is to remain undetected and to avoid terrain that is likely to attract heavy artillery fire. Positions should be as well protected against artillery as possible. Move, if possible, when positions come under artillery fire.

c. **Helicopters.** Attack helicopters routinely follow artillery preparation. Again, ATGMs are high-priority targets.

d. **Direct Fire.** Enemy forces will attempt to suppress or destroy friendly ATGM systems with direct fire.

e. **Grenade Launcher.** This weapon is very effective against a tripod- or HMMWV-mounted TOW, and is somewhat effective against lightly armored vehicles such as the Bradley.

f. **Lasers.** Most armies use laser range finders and target designators. Many of these are quite powerful and can cause temporary or permanent damage to a gunner's eyes when he is using direct-view magnifying optics. Some of these devices can even damage the sights themselves. Thermal sights are less vulnerable to lasers. Laser weapons, such as the PRC-produced ZM-87, or even frequency agile laser weapons may be encountered on the future battlefield. Many countries, such as China, Russia, and France, have extensive laser development and procurement programs.

g. **Battlefield Obscurants.** The AN/TAS-4 series of thermal sights is able to penetrate many kinds of battlefield obscurants, but not all. Bispectral smoke and battlefield dust, among others, are largely opaque to thermal sights if they are thick enough. In addition, when using a basic or ITOW, the target may be seen, but the system is unable to guide the missile through the obscurants.

h. **Combined Arms.** Enemy doctrine prefers to fight mounted, but if they run into strong ATGM defenses, they will dismount their infantry and advance at foot pace, with the infantry engaging personnel and ATGMs. The tanks will advance first, at about 6 kilometers per hour, and the infantry will closely follow the tanks. The fighting vehicles will follow the infantry and provide supporting fire.

H-6. INFANTRY ANTITARMOR COUNTER-COUNTERMEASURES

United States forces must have counter-countermeasures against reconnaissance, artillery, helicopters, direct fire, grenade launchers, lasers, obscurants, and combined arms.

a. **Reconnaissance.** The best counter to reconnaissance elements is effective counterreconnaissance operations by defending forces. Equally important is effective and rigidly enforced camouflage and concealment.

b. **Artillery.** The best defense against artillery is not to be discovered by the enemy's reconnaissance elements. This may not be possible, so positions must be as well prepared against heavy artillery fire as time permits. When all else fails, it may be necessary to move positions. Secondary and alternate positions should be preselected for this eventuality.

c. **Helicopters.** The best defense against helicopters is not to be spotted in the first place. TOW positions should only engage helicopters in self-defense when they are already spotted and under attack. Helicopters not engaging TOW positions should be left to regular ADA assets. (See Chapter 5 for more detailed information.)

d. **Direct Fire.** TOW squads must maximize their chances of winning engagements with tanks and fighting vehicles by choosing positions with good cover and concealment, striving for flank shots, striving for surprise, exercising good fire control, and carefully prioritizing their targets, to ensure that targets posing the most immediate threat are engaged first. This is, in fact, precisely the type of combat the TOW was intended to engage in.

e. **Grenade Launcher.** TOW squads must try to choose positions where it is difficult for enemy grenade launcher teams to get in range without being detected and engaged. Where the terrain is too close to permit this, TOW squads must emplace behind friendly infantry, and let the infantry engage the grenade launcher team.

f. **Lasers.** Protective lenses should be installed on optical devices. The protective covers or doors on these devices should be closed when protective lenses are not available. Soldiers should wear laser-proof goggles at all times. (Soldiers are vulnerable to eye damage even when not looking through sighting devices.)

g. **Battlefield Obscurants.** TOW crews should train under realistic battlefield obscurant conditions whenever possible, so that they better understand the limitations of their thermal sights. Alternate positions that view the battlefield from a different angle and that may afford a better view should be identified and prepared. The battlefield should be carefully examined and marked to give gunners a better perspective of where things are when the field becomes partially or largely obscured. Enemy smokescreens normally end about 1,000 meters from our positions, so gunners should be prepared to engage targets immediately when they emerge from the smokescreen.

h. **Combined Arms.** TOW positions are ill-equipped to defend themselves against dismounted infantry. To survive a properly executed combined arms attack, TOW squads should be integrated into a combined arms defense that includes overwatching dug-in infantry, fighting vehicles, and tanks. The TOW was never intended as a stand-alone weapon system, and leaders who attempt to employ it as such guarantee their destruction.

H-7. DIRECTED-ENERGY WEAPONS

Directed-energy weapons include lasers, microwave radiation emitters, and particle beam generators. By far the most likely type that ground troops will see in combat is the laser. This can take the form of a powerful laser range finder used as a weapon or as a specially designed laser weapon. The common perception of the laser is as a science fiction weapon that can vaporize a tank with a single blast. This is far from the truth. In fact, the laser beam will normally be invisible and is generally able only to temporarily blind a soldier who is hit in the eyes with it. If the soldier is using an optical sight of some kind, the effect is magnified. More powerful lasers may cause permanent blindness, but these are rare. Some lasers are also powerful enough to damage the lenses of optical sights or the electronics of electrooptical devices (generally thermal sights and image intensifiers). (For means of defense, see paragraph H-4 of this appendix. See FM 7-91, Appendix F for a more detailed discussion.)

H-8. IMPROVED ARMOR TECHNOLOGY

In the past decade, there has been a revolution in armor technology. Research and new developments have come from Europe, the United States, and Israel. These improvements are continuing worldwide, and the antiarmor crewman must keep up with them. These improvements are also becoming much more common in third world armies, so the antiarmor crewman must expect to meet them wherever he goes. In addition, many older tanks and other AFVs are being retrofitted with improved armor protection, so it is not just the newer vehicles we need be concerned with. These advanced armor configurations improve the vehicles' survivability against all weapons, but for the most part they are specifically designed to protect against HEAT warheads. This is a complex technical subject, and growing more so every year, but essentially these improved types of armor fall into four categories.

a. **Reactive Armor**. Originally fielded by the Israeli Army, reactive armor appeared shortly after on enemy tanks. Reactive armor comes in several varieties, but the principle is essentially the same on all. The armor consists of blocks of explosives sandwiched between two metal plates and bolted on the outside of the vehicle. Small-arms and artillery shrapnel will not set off the blocks. However, when a HEAT round strikes the block, the explosive ignites and blows outwards. The blast and the moving steel plates disperse and deflect the jet of the HEAT warhead, dramatically reducing its ability to penetrate armor. Many countries are now fielding different versions of reactive armor. One of its advantages is that it can easily be retrofitted onto older vehicles.

b. **Laminated Armor.** Laminated armor was developed in the West, most specifically by England, the United States, and West Germany. It consists of flat layers of steel armor plate with layers of ceramics, fiberglass, or other nonmetallic materials in between. This armor is highly effective against all types of weapons, but is difficult and expensive to manufacture. Vehicles with laminated armor are characterized by flat, slab sides, such as on the M1 and Leopard II.

c. **Composite Armor.** Composite armor consists of a nonmetallic core (usually some kind of ceramic) around which the rest of the steel of the hull or, more commonly, the turret, is molded. This is much more effective than conventional steel armor against all types of weapons, but less so than laminated armor. However, it is less difficult and expensive to manufacture. Hulls made of composite armor do not have to be slab sided, like those made of laminated armor.

d. **Applique Armor.** Applique armor is essentially extra plates mounted or welded on top of the hull or turret of a vehicle. They can be made of any material, but are frequently made of ceramic or laminated materials. Like reactive armor, applique armor is an easy and cost-effective way of improving the protection of older vehicles.

H-9. COUNTERMEASURES TO IMPROVED ARMOR TECHNOLOGY

The long-term answer to improved armor technology is primarily a technical issue: improving the kill mechanisms of our missiles to defeat the improved armor.

a. Two responses to improved armor have already been fielded: the TOW 2A and the TOW 2B. Other improved kill mechanisms will be fielded as time passes.

(1) The TOW 2A has an explosive charge in the tip of the extensible probe to prematurely detonate reactive armor and minimize its effect on the penetration abilities of the missile.

(2) The TOW 2B employs a new kill mechanism. It flies about 6 or 7 feet above the line of sight and detonates above the hull or turret of the tank. When it detonates, it creates two explosively formed penetrators (EFPs), which are metallic slugs travelling at very high speeds. These EFPs strike the top of the vehicle where the armor is thinnest and there are fewer reactive armor plates.

b. TOW crews can expect to be issued a mix of TOW missile types on the battlefield, with widely varying capabilities. Gunners and leaders must be familiar with the different missile types and their respective capabilities. The proper type of missile must be chosen for each type of target (Table H-2).

c. TOW crews must strive harder than ever to find positions where they can engage enemy vehicles from the flank. Modern tanks with reactive armor have become increasingly difficult to kill from the front.

THREAT VEHICLE TYPE TARGETS	SELECTION PRIORITY			
	FIRST	SECOND	THIRD	FOURTH
Tanks with applique armor.	TOW 2B	TOW 2A	TOW 2	ITOW
Tanks with explosive reactive armor.	TOW 2B	TOW 2A	TOW 2	ITOW
Tanks without applique/ reactive armor.	TOW 2B	TOW 2A	TOW 2	ITOW
Light armored personnel carriers.	TOW 2	TOW 2A	TOW 2B	ITOW
Light armored wheeled vehicles.	TOW 2	TOW 2A	TOW 2B	ITOW
Antiaircraft vehicles.	TOW 2	TOW 2A	TOW 2B	ITOW
Armored vehicles in hull defilade positions.	TOW 2B	TOW 2A	TOW 2	ITOW
Bunkers/fortifications.	TOW 2	TOW 2A	ITOW	TOW 2B

Table H-2. Missile selection priority chart.

GLOSSARY

ACCP	Army Correspondence Course Program
ADA	air defense artillery
AFV	armored fighting vehicle
ANCOC	Advanced noncommissioned officer course
APC	armored personnel carrier
AR	Army regulation
ARTEP	Army training and evaluation program
ATEA	anticipated target engagement area
ATGM	antitank guided missile
ATTN	attention
ATWESS	antitank weapon effect signature simulator
AZ	azimuth
BFV	Bradley fighting vehicle
BNCOC	Basic noncommissioned officer course
BPC	battery power condition
CO	company
cm	centimeter
CVC	combat vehicle crewman (helmet)
CVKI	combat vehicle kill indicator
DA	Department of the Army
DS	direct support
EFP	explosively formed projectiles
ELEV	elevation
EMER PWR	emergency power
EOD	explosive ordnance disposal
FIST	fire support team
FM	field manual
FO	forward observer
FSO	fire support officer
FTX	field training exercise

GCP	gunner's control panel
GHC	gunner's hand control
GST	Gunner's Skill Test
HMMWV	high-mobility, multipurpose wheeled vehicle
IAW	in accordance with
IFV	infantry fighting vehicle
IMT	infantry moving target
IR	infrared
ITA	image transfer assembly
ITOW	improved TOW
IVD	interactive video disk
km	kilometers
km/h	kilometers per hour
LOS	line of sight
LTD	laser target detector
LTID	laser target interface device
MCLOS	manual command to line-of-sight guidance
METL	mission-essential task list
METT-T	mission, enemy, terrain, troops, and time available
MGS	missile guidance set
MILES	Multiple Integrated Laser Engagement System
MOPP	mission-oriented protective posture
MOS	military occupation specialty
MPRC	multipurpose range complex
MSR	missile simulation round
MV	moving
MWLD	man-worn laser detectors
MWO	modified work order
NATO	North Atlantic Treaty Organization
NBC	nuclear, biological, and chemical
NCO	noncommissioned officer
NTC	National Training Center

OCL	operator's checklist
OPFOR	opposing force
OSUT	one station unit training
PGTS	Precision Gunnery Training System
PLT	platoon
PMCS	preventive maintenance checks and services
psi	pounds per square inch
ROW	rest of the world
rpm	revolutions per minute
SACLOS	semiautomatic command-to-line-of-sight guidance
SDZ	surface danger zone
SFC	sergeant first class
SGT(P)	sergeant (promotable)
SL	skill level; squad leader (figures only)
SLP	squad leader's periscope
SM	soldier's manual
SOP	standing operating procedure
SQD	squad
ST	stationary
STBY	standby
STRAC	Standards in Training Commission
STP	soldier training publication
STX	situational training exercise
TFTT	TOW Field Tactical Trainer
TG	trainer's guide
TGT	TOW Gunnery Trainer
TOW	tube-launched, optically tracked, wire-guided
TM	technical manual (training manual)
TRP	target reference point
TVPC	TOW vehicle power conditioner
US	United States
VDC	volts direct current
VPC	vehicle power conditioner

REFERENCES

DOCUMENTS NEEDED

These documents must be available to the intended users of this publication.

AR 75-1	Malfunctions Involving Ammunition and Explosives. 23 April 2001.
ARTEP 7-91-Drill	Drills for the Antiarmor (TOW) Platoon, Section, and Squad. 26 July 2002.
ARTEP 7-91-MTP	Mission Training Plan for Antiarmor Company/Platoon/Section. 12 October 1989.
DA Form 7331-R	Scorecard for Gunner's Skill Test-Part 1. August 2003.
DA Form 7332-R	Scorecard for Gunner's Skill Test-Part 2-Carrier-Specific Tasks. August 2003.
DA Form 7333-R	TOW Gunnery Tables 1 and 2: Individual Gunnery Practice and Qualification Scorecard. August 2003.
DA Form 7334-R	TOW Gunnery Tables 3 and 4: Advanced Gunnery Practice and Qualification Scorecard. August 2003.
DA Form 7335-R	TOW Gunnery Tables 5 and 6: Baseline TFTT Gunnery Practice and Qualification Scorecard. August 2003.
DA Form 7336-R	TOW Gunnery Table 7: Squad Gunnery Practice Scoresheet. August 2003.
DA Form 7337-R	TOW Gunnery Table 8: Squad Gunnery Qualification Scoresheet. August 2003.
DA Form 7338-R	TOW Gunnery Table 9: Section Gunnery Practice Scoresheet. August 2003.
DA Form 7339-R	Tow Gunnery Table 10: Section Gunnery Qualification Scoresheet. August 2003.

DA Form 7340-R	TOW Gunnery Table 11: Platoon Gunnery Practice Scoresheet. August 2003.
DA Form 7341-R	TOW Gunnery Table 12: Platoon Gunnery Qualification Scoresheet. August 2003.
DA Pam 350-38	Standards in Weapons Training. 01 October 2002.
DA Pam 385-63	Range Safety. 10 April 2003.
DA Pam 611-21	Military Occupational Classification and Structure. 31 March 1999.
FM 3-21.91(FM 7-91)	Tactical Employment of Antiarmor Platoons, Companies, and Battalions. 26 November 2002.
FM 3-37.3(FM 3-5)/ with change	NBC Decontamination. 28 July 2000.
FM 5-250	Explosives and Demolitions. 30 July 1998.
FM 10-500-29	Airdrop of Supplies and Equipment: Rigging Heavy Antitank Assault Weapon System (TOW). 14 April 2000.
FM 25-101	Battle Focused Training. 30 September 1990.
GTA 17-02-011	Combat Vehicle Identification (CVI) Training Cards. 1984.
GTA 17-02-013	Armored Vehicle Recognition. 01 January 1997.
STP 7-11BCHM24-SM-TG	Soldier's Manual and Trainer's Guide for MOS 11B, 11C, 11H, and 11M Infantry Skill Levels 2/3/4. 19 June 2000.
TB MED 524	Occupational and Environmental Health: Control of Hazards to Health From Laser Radiation. 20 June 1985.
TC 25-8/with change	Training Ranges. 25 February 1992.

TM 9-1265-368-10-2	Operator's Manual for Multiple Integrated Laser Engagement System (MILES), Simulator System, Firing Laser: M64 (NSN 1265-01-077-6083) for TOW Standard, TOW Long Range, TOW II, TOW II Extended Weapon System. 15 July 1988.
TM 9-1410-470-34	Direct Support and General Support Maintenance Manual for Surveillance of Guided Missile Surface Attack, BGM-71A, BGM-71A-1, BGM-71A-2, BGM-71A-2A, BGM-71A-3, BGM-71A-3A, BGM-71C, BGM-71C-1, BGM-71C-1A, BGM-71D, or BGM-71E and Guided Missile Practice BTM-71A, BTM-71A-1, BTM-71A-2, BTM-71A-2A, BTM-71A-3, or BTM-71A-3A (TOW Heavy Antitank/Assault Weapon System). 01 March 1988.
TM 9-1425-450-12	Operator and Organizational Maintenance Manual for TOW 2 Weapon System, Guided Missile System M220A2 (NSN 1440-014-9034) (TOW 2 Heavy Antitank/Assault Weapon System). 25 May 1983.
TM 9-2320-280-10	Operator's Manual for Truck, Utility: TOW Carrier, Armored, 1-1/4 Ton, 4X4, M966. 31 January 1996.
TM 9-6920-452-10	Trainer, Gunnery, Antitank: AN/TWQ-T1 (TOW Gunnery Trainer). 01 August 1991.
TM 9-6920-453-10	Trainer, Guided Missile: AN/TWQ-T3 (TOW Field Tactical Trainer). 20 March 1991.
TM 750-244-6	Procedures for Destruction of Tank-Automotive Equipment to Prevent Enemy Use (US Army Tank-Automotive Command). 03 October 1972.

INTERNET WEB SITES

Army Publishing Directorate
http://www.usapa.army.mil

Reimer Digital Library
http://www.adtdl.army.mil

INDEX

GUNNER'S SKILL TEST - PART 1

For use of this form, see FM 3-22.34. The proponent agency is TRADOC.

NAME	RANK	UNIT	DATE

TEST#	TASK NUMBER	TASK TITLE	GO/NO GO
1	071-056-0004	MAINTAIN AN M220-SERIES LAUNCHER SYSTEM	
2	071-056-0007	LOAD AN M220-SERIES LAUNCHER SYSTEM	
3	071-035-0008	UNLOAD AN M220-SERIES LAUNCHER SYSTEM	
4	071-056-0009	ENGAGE TARGETS WITH AN M220-SERIES LAUNCHER SYSTEM	
5	071-056-0010	PERFORM IMMEDIATE ACTION FOR AN M220-SERIES LAUNCHER SYSTEM	
6	071-056-0005	CONDUCT A SYSTEM CHECK OUT ON AN M220A1 LAUNCHER SYSTEM	
7	071-056-0013	CONDUCT A SYSTEM CHECK OUT ON AN M220A2 LAUNCHER SYSTEM	
8	071-056-0030	DETERMINE M220-SERIES LAUNCHER SYSTEM FIRING LIMITATIONS	
9	NO TASK NUMBER	IDENTIFY COMBAT VEHICLES	
10	NO TASK NUMBER	DETERMINE IF A TARGET CAN BE ENGAGED BY AN M220-SERIES LAUNCHER SYSTEM (SEE FM 3-22.34, CH5)	
11	071-317-0000	PREPARE AN ANTIARMOR RANGE CARD	
12	NO TASK NUMBER	ASSEMBLE AN M220-SERIES LAUNCHER SYSTEM	

DA FORM 7331-R, AUG 2003

APD V1.00

GUNNER'S SKILL TEST - PART 2						
For use of this form, see FM 3-22.34. The proponent agency is TRADOC.						
NAME			RANK	UNIT		DATE
TEST#	TASK NUMBER	TASK TITLE				GO/NO GO
1	NO TASK NUMBER	PLACE THE M966 VEHICLE IN THE READY-TO-FIRE CONFIGURATION (SEE FM 3-22.34 CH 3)				
2	071-056-9932	CONDUCT DISMOUNTING AND REMOUNTING OF AN M220A1 LAUNCHER SYSTEM ON AN M966 VEHICLE (M220A1- EQUIPPED UNITS ONLY)				
3	071-056-0034	CONDUCT DISMOUNTING AND REMOUNTING OF AN M220A2 LAUNCHER SYSTEM ON AN M966 VEHICLE (M220A2 - EQUIPPED UNITS ONLY)				
4	071-034-0001	LOAD AN M243 OR M259 SMOKE GRENADE LAUNCHER				
5	071-034-0002	UNLOAD AN M243 OR M259 SMOKE GRENADE LAUNCHER				

DA FORM 7332-R, AUG 2003 APD V1.00

TOW GUNNERY TABLES 1 AND 2: INDIVIDUAL GUNNERY

☐ PRACTICE ☐ QUALIFICATION

For use of this form, see FM 3-22.34. The proponent agency is TRADOC.

DATA REQUIRED BY PRIVACY ACT OF 1974

AUTHORITY: 10USC 30129g. Executive Order 9397. **PRINCIPAL PURPOSE(S):** Records individual's performance on record fire range. **ROUTINE USE(S):** Evaluation of individual's proficiency and basis for determination of award of proficiency badge; SSN is used for positive identification purposes only. **MANDATORY OR VOLUNTARY DISCLOSURE AND EFFECT ON INDIVIDUAL NOT PROVIDING INFORMATION:** Voluntary; individuals not providing information cannot be rated/scored on a mass basis.

NAME	RANK	SSN	UNIT

DATE	DISK#

MISSION#	INDIVIDUAL GUNNERY TASK TESTED	SIGHT	SCORE	SUCCESSFUL	
#				YES	NO
#				YES	NO
#				YES	NO
#				YES	NO
#				YES	NO
#				YES	NO
#				YES	NO
#				YES	NO
#				YES	NO
#				YES	NO
	TOTAL TRACKING SCORE			SUCCESSFUL MISSIONS: ____	

GUNNER CLASSIFICATION

EXPERT / 875 - 1,000 1ST CLASS / 750 - 874 2ND CLASS / 625 - 749

SCORING INDIVIDUAL GUNNERY TABLES

IN THE INDIVIDUAL GUNNERY TABLES, THE GUNNER'S RATING IS DETERMINED BY HIS TOTAL POINT SCORE FOR 10 MISSIONS WITH AT LEAST 6 SUCCESSFUL MISSIONS. A SUCCESSFUL MISSION IS ONE WHERE THE GUNNER CORRECTLY FOLLOWS THE TRAINERS FIRE COMMANDS AND ENGAGES AND DESTROYS THE CORRECT TARGET.

TOW GUNNERY TABLES 3 AND 4: ADVANCED GUNNERY

☐ **PRACTICE** ☐ **QUALIFICATION**

For use of this form, see FM 3-22.34. The proponent agency is TRADOC.

NAME	RANK	SSN	UNIT

DATE	DISK#

MISSION#	INDIVIDUAL GUNNERY TASK TESTED	SIGHT	SCORE	SUCCESSFUL	
#				YES	NO
#				YES	NO
#				YES	NO
#				YES	NO
#				YES	NO
#				YES	NO
#				YES	NO
#				YES	NO
#				YES	NO
#				YES	NO
	TOTAL TRACKING SCORE			SUCCESSFUL MISSIONS: ______	

GUNNER CLASSIFICATION

EXPERT / 875 - 1,000 1ST CLASS / 750 - 874 2ND CLASS / 625 - 749

SCORING ADVANCE GUNNERY TABLES

IN THE INDIVIDUAL GUNNERY TABLES, THE GUNNER RATING IS DETERMINED BY HIS TOTAL POINT SCORE FOR 10 MISSIONS, PLUS HE MUST HAVE AT LEAST 6 SUCCESSFUL MISSIONS. A SUCCESSFUL MISSION IS ONE WHERE THE GUNNER CORRECTLY FOLLOWS THE TRAINERS FIRE COMMANDS AND ENGAGES AND DESTROYS THE CORRECT TARGET. IF THE MISSION IS A FRATRICIDE MISSION, THE GUNNER SHOULD **NOT** ENGAGE THE TARGET. IF HE **DOES NOT** ENGAGE THE FRIENDLY VEHICLE, HE SCORES 100 POINTS AND A SUCCESSFUL MISSION. IF HE **DOES** ENGAGE THE FRIENDLY VEHICLE, HE SCORES 0 POINTS AND AN UNSUCCESSFUL MISSION. IN TARGET ENGAGEABILITY AND EVASIVE TARGET SCENARIOS IT DOES NOT MATTER WHICH TARGET IS ENGAGED.

TOW GUNNERY TABLES 5 AND 6: BASELINE TFTT GUNNERY

☐ PRACTICE ☐ QUALIFICATION

For use of this form, see FM 3-22.34. The proponent agency is TRADOC.

SQUAD	SECTION	PLATOON	COMPANY

DATE	DAY	NIGHT

ENGAGEMENT	TARGET DESCRIPTION	HIT/MISS/TIME	GO/NG GO
TARGET 1	STATIONARY T72/FLANK/ 1,000 - 1,500 METERS	HIT MISS TIME:	
TARGET 2	STATIONARY T72/FLANK/ 2,100 - 3,750 METERS	HIT MISS TIME:	
TARGET 3	STATIONARY T72/FRONTAL/ 2,100 - 3,750 METERS	HIT MISS TIME:	
TARGET 4	STATIONARY T72/OBLIQUE/ 2,100 - 3,750 METERS	HIT MISS TIME:	
TARGET 5	STATIONARY T72/DEFILADE/ 800 - 1,500 METERS/NBC	HIT MISS TIME:	
TARGET 6	MOVING T72/FLANK/ L-R, 30 KPH 1,600 - 2,000 METERS/NBC	HIT MISS TIME:	
TARGET 7	MOVING T72/FLANK/ L-R, 30 KPH 1,600 - 2,000 METERS/NBC	HIT MISS TIME:	
TARGET 8	MOVING T72/FLANK/ L-R, 5 KPH 1,000 - 1,500 METERS	HIT MISS TIME:	
TARGET 9	MOVING T72/FLANK/ L-R, 10 KPH 2,100 - 3,750 METERS	HIT MISS TIME:	
TARGET 10	MOVING T72/FLANK/ L-R, 40 KPH 1,000 - 1,500 METERS	HIT MISS TIME:	
		OVERALL GO/NO GO:	

TO RECEIVE A "GO" ON ANY TARGET ENGAGEMENT, THE GUNNER MUST HIT THE TARGET WITHIN 30 SECONDS AFTER THE TARGET IS EXPOSED. TO RECEIVE A "GO" ON TABLE 5, THE CREW MUST RECEIVE A "GO" ON "INSTALL TFTT" AND A "GO" ON AT LEAST 6 OUT OF 10 TARGET ENGAGEMENTS, BOTH DAY AND NIGHT.

TOW GUNNERY TABLE 7: SQUAD GUNNERY PRACTICE

For use of this form, see FM 3-22.34. The proponent agency is TRADOC.

TC: ________ LDR: ________ GNR: ________ DVR: ________ DATE: ________ TIME: ________

PLATOON: ________ COMPANY: ________ BATTALION: ________

TASK	CONDITIONS/ TARGETS/ SITUATION	TOTAL TARGET EXPOSURE TIME	HITS	ENGAGEMENT TIMES	POINTS	CREW/ LDR CUTS	TOTAL POINTS
1. ENGAGE A SINGLE TARGET	FIRED FROM BASELINE. 3,000 - 3,750 METERS 1 STATIONARY TANK	TFTT 25 SECS ALL VEHICLES	1				
2. ENGAGE MULTIPLE TARGETS	FIRED FROM PHASE LINE 2. 2,000 - 2,500 METERS 1 - ST TANK, 1 - MV TANK, 1 - MV BMP	TFTT HMMWV/2:30	1 2 3				
3. ENGAGE MULTIPLE TARGETS	FIRED FROM PHASE LINE 3. 2,500 - 3,000 METERS 1 - MV TANK, 1 - ST BMP	TFTT HMMWV/1:40	1 2				
4. ENGAGE MULTIPLE TARGETS	FIRED FROM PHASE LINE 2. 2,500 - 3,000 METERS 1 ST TANK, 1 - MV TANK (NBC)	TFTT HMMWV/1:40	1 2				
5. ENGAGE MULTIPLE TARGETS	FIRED FROM BASELINE. 2,500 - 3,000 METERS 1 - MV TANK, 1 - MV BMP (NBC)	TFTT HMMWV/1:40	1 2				
					TOTAL SCORE		

NOTES:

1. ENGAGEMENT TIMES ARE CALCULATED FROM THE TIME THE TARGET IS EXPOSED TO THE TIME THE TARGET IS HIT. USE THE LIVE-FIRE POINT CALCULATION SHEET FOR SECOND AND THIRD ENGAGEMENTS FROM THE SAME POSITION. THE TIME BEGINS WHEN THE VEHICLE REOCCUPIES ITS FIRING POSITION (STOPS MOVING) AND ENDS WHEN THE TARGET IS HIT.
2. EACH TASK IS WORTH 100 POINTS. EACH TARGET IS ALSO WORTH 100 POINTS FOR TASKS WITH MULTIPLE ENGAGEMENTS. TOTAL THE NUMBER OF POINTS AND DIVIDE BY THE NUMBER OF TARGETS PRESENTED.
3. CREW/LEADER CUTS: SEE FM 3-22.34, CH 4.
4. SCORING TABLE 7: TABLE 7 IS PERFORMED TWICE, ONCE DURING THE DAY AND ONCE AT NIGHT. EACH EVENT IS WORTH A MAX OF 500 POINTS. ADD THE TWO SCORES TOGETHER FOR A POSSIBLE TOTAL OF 1,000 POINTS. SQUADS MUST ACHIEVE A MINIMUM TOTAL OF 700 POINTS TO VERIFY ON TABLE 7 AND PROGRESS TO TABLE 8.

DA FORM 7336-R, AUG 2003 APD V1.00

TOW GUNNERY TABLE 8: SQUAD GUNNERY QUALIFICATION

For use of this form, see FM 3-22.34. The proponent agency is TRADOC.

TC: ____________ LDR: ____________ GNR: ____________ DVR: ____________ DATE: ____________ TIME: ____________

PLATOON: ____________ COMPANY: ____________ BATTALION: ____________

TASK	CONDITIONS/ TARGETS/ SITUATION	TOTAL TARGET EXPOSURE TIME	HITS	ENGAGEMENT TIMES	POINTS	CREW/ LDR CUTS	TOTAL POINTS
1. ENGAGE A SINGLE TARGET	FIRED FROM BASELINE. 1,500 - 2,000 METERS 1 STATIONARY TANK	TFTT 25 SECS ALL VEHICLES	1				
2. ENGAGE MULTIPLE TARGETS	FIRED FROM PHASE LINE 1. 2,000 - 2,500 METERS 1 - MV TANK, 1 - ST BMP	TFTT HMMWV/1:40	1 2				
3. ENGAGE MULTIPLE TARGETS	FIRED FROM PHASE LINE 2. 3,000 - 3,750 METERS 2 - MV TANK, 1 - ST BMP	TFTT HMMWV/2:45	1 2 3				
4. ENGAGE MULTIPLE TARGETS	FIRED FROM PHASE LINE 1. 1,500 - 2,000 METERS 1 - ST TANK, 1 - MV TANK (NBC)	TFTT HMMWV/1:40	1 2				
5. ENGAGE MULTIPLE TARGETS	FIRED FROM BASELINE. 2,500 - 3,000 METERS 1 - MV TANK, 1 - MV BMP (NBC)	TFTT HMMWV/1:40	1 2				
					TOTAL SCORE		

NOTES:

1. TABLE 8 IS CONDUCTED AND TIMED IN EXACTLY THE SAME MANNER AS TABLE 7.
2. TABLE 8, UNLIKE TABLE 7, IS FOR QUALIFICATION. THE SQUAD IS RATED ON ITS FINAL SCORE, RATHER THAN SIMPLY HAVING TO MEET A MINIMUM STANDARD.
3. CREW/LEADER CUTS: SEE FM 3-22.34, CH 4.
4. SCORING TABLE 8: TABLE 8 IS PERFORMED TWICE, ONCE DURING THE DAY AND ONCE AT NIGHT. EACH EVENT IS WORTH A MAX OF 500 POINTS. ADD THE TWO SCORES TOGETHER FOR A POSSIBLE TOTAL OF 1,000 POINTS. SQUADS ARE RATED ON THE FOLLOWING BASIS:

DISTINGUISHED: 900 - 1,000 POINTS SUPERIOR: 800 - 899 POINTS QUALIFIED: 700 - 799 POINTS UNQUALIFIED: 0 - 699 POINTS

DA FORM 7337-R, AUG 2003

APD V1.00

TOW GUNNERY TABLE 9: SECTION GUNNERY PRACTICE

For use of this form, see FM 3-22.34. The proponent agency is TRADOC.

SECTION TC: ______ LDR: ______ GNR: ______ DVR: ______ DATE: ______

SQUAD TC: ______ LDR: ______ GNR: ______ DVR: ______ TIME: ______

PLATOON: ______ COMPANY: ______ BATTALION: ______

TASK	CONDITIONS/ TARGETS/ SITUATION	TOTAL TARGET EXPOSURE TIME	HITS	ENGAGEMENT TIMES	POINTS	CREW/ LDR CUTS	TOTAL POINTS
1. ENGAGE A SINGLE TARGET	FIRED FROM BASELINE. 3,000 - 3,750 METERS 1 - ST TANK, 1 - ST BMP	TFTT 40 SECS ALL VEHICLES	1 2				
2. ENGAGE MULTIPLE TARGETS	FIRED FROM PHASE LINE 2. 2,000 - 2,750 METERS 1- ST BMP	TFTT 25 SECS ALL VEHICLES	1				
3. ENGAGE MULTIPLE TARGETS	FIRED FROM PHASE LINE 3. 1,000 - 1,500 METERS 2 - MV BMPS, 1 - ST TANK	TFTT HMMWV/1:40	1 2 3				
4. ENGAGE MULTIPLE TARGETS	FIRED FROM PHASE LINE 2. 2,500 - 3,000 METERS 2 - MV TANKS, 2 - ST TANKS (NBC)	TFTT HMMWV/2:38	1 2 3 4				
5. ENGAGE MULTIPLE TARGETS	FIRED FROM BASELINE. 1,000 - 1,500 METERS 1 - ST TANK, 1 - ST BMP (NBC)	TFTT 25 SECS ALL VEHICLES	1 2				
					TOTAL SCORE		

NOTES:

1. TABLE 9 IS CONDUCTED AND TIMED IN EXACTLY THE SAME MANNER AS TABLES 7 AND 8.

2. SCORING TABLE 9: TABLE 9 IS PERFORMED TWICE, ONCE DURING THE DAY AND ONCE AT NIGHT. EACH EVENT IS WORTH A MAX OF 500 POINTS. ADD THE TWO SCORES TOGETHER FOR A POSSIBLE TOTAL OF 1,000 POINTS. TABLE 9 IS A PRACTICE TABLE. TO VERIFY ON TABLE 9 AND PROGRESS TO TABLE 10, THE SECTION MUST ACHIEVE A MINIMUM COMBINED SCORE OF AT LEAST 700 POINTS, AFTER CREW AND LEADER CUTS.

TOW GUNNERY TABLE 10: SECTION GUNNERY QUALIFICATION

For use of this form, see FM 3-22.34. The proponent agency is TRADOC.

SECTION TC: ____ LDR: ____ GNR: ____ DVR: ____ DATE: ____

SQUAD TC: ____ LDR: ____ GNR: ____ DVR: ____ TIME: ____

PLATOON: ____ COMPANY: ____ BATTALION: ____

TASK	CONDITIONS/ TARGETS/ SITUATION	TOTAL TARGET EXPOSURE TIME	HITS	ENGAGEMENT TIMES	POINTS	CREW/ LDR CUTS	TOTAL POINTS
1. ENGAGE A SINGLE TARGET	FIRED FROM BASELINE. 3,000 - 3,750 METERS 2 - MV TANK, 2 - ST BMPS	TFTT HMMWV/2:38	1 2 3 4				
2. ENGAGE MULTIPLE TARGETS	FIRED FROM PHASE LINE 2. 2,000 - 2,500 METERS 1- MV TANK	TFTT 40 SECS ALL VEHICLES	1				
3. ENGAGE MULTIPLE TARGETS	FIRED FROM PHASE LINE 3. 1,000 - 1,500 METERS 1 - MV BMP, 1 - ST TANK	TFTT 25 SECS ALL VEHICLES	1 2				
4. ENGAGE MULTIPLE TARGETS	FIRED FROM PHASE LINE 2. 2,500 - 3,000 METERS 1 - ST BMP, 2 - MV TANKS (NBC)	TFTT HMMWV/2:40	1 2 3				
5. ENGAGE MULTIPLE TARGETS	FIRED FROM BASELINE. 3,000 - 3,750 METERS 1 - ST TANK, 1 - ST BMP (NBC)	TFTT 25 SECS ALL VEHICLES	1 2				
					TOTAL SCORE		

NOTES:

1. TABLE 10 IS CONDUCTED AND TIMED IN EXACTLY THE SAME MANNER AS TABLE 8 AND 9.
2. SCORING TABLE 10: TABLE 10 IS PERFORMED TWICE, ONCE DURING THE DAY AND ONCE AT NIGHT. EACH EVENT IS WORTH A MAX OF 500 POINTS. ADD THE TWO SCORES TOGETHER FOR A POSSIBLE TOTAL OF 1,000 POINTS. TABLE 10 IS A QUALIFICATION TABLE. TO PROGRESS TO TABLE 11, THE SECTION MUST ACHIEVE A MINIMUM COMBINED SCORE OF AT LEAST 700 POINTS.

DA FORM 7339-R, AUG 2003

APD V1.00

TOW GUNNERY TABLE 11: PLATOON GUNNERY PRACTICE

For use of this form, see FM 3-22.34. The proponent agency is TRADOC.

SECTION 1:	TOW CREW 1:	TC: ____	LDR: ____	GNR: ____	DVR: ____	DATE: ____
	TOW CREW 2:	TC: ____	LDR: ____	GNR: ____	DVR: ____	
SECTION 2:	TOW CREW 1:	TC: ____	LDR: ____	GNR: ____	DVR: ____	
	TOW CREW 2:	TC: ____	LDR: ____	GNR: ____	DVR: ____	TIME: ____
SECTION 3:	TOW CREW 1:	TC: ____	LDR: ____	GNR: ____	DVR: ____	
	TOW CREW 2:	TC: ____	LDR: ____	GNR: ____	DVR: ____	

PLATOON: ____ COMPANY: ____ BATTALION: ____

TASK	CONDITIONS/ TARGETS/ SITUATION	TOTAL TARGET EXPOSURE TIME	HITS	ENGAGEMENT TIMES	POINTS	CREW/ LDR CUTS	TOTAL POINTS
1. ENGAGE A SINGLE TARGET	FIRED FROM BASELINE. 3,000 - 3,750 METERS 4 - MV TANKS, 2 - ST BMPS	TFTT HMMWV/3:26	1 2 3 4 5 6				
2. ENGAGE MULTIPLE TARGETS	FIRED FROM PHASE LINE 2. 2,000 - 3,000 METERS 4 - MV BMPS 2 - ST BMPS	TFTT HMMWV/2:26	1 2 3 4 5 6				
3. ENGAGE MULTIPLE TARGETS	FIRED FROM PHASE LINE 3. 1,000 - 2,000 METERS 2 - ST TANKS, 2 - ST BMPS	TFTT 25 SECS ALL VEHICLES	1 2 3 4				
4. ENGAGE MULTIPLE TARGETS	FIRED FROM PHASE LINE 2. 2,000 - 3,000 METERS 2- MV TANKS, 2 - MV BMPS (NBC)	TFTT BMP/43 SEC TANK/1:03	1 2 3 4				
5. ENGAGE MULTIPLE TARGETS	FIRED FROM BASELINE. 3,000 - 3,750 METERS 2 - ST TANKS, 2 - ST BMPS (NBC)	TFTT 25 SECS ALL VEHICLES	1 2 3 4				
					TOTAL SCORE		

NOTES:

1. TABLE 11 IS CONDUCTED, TIMED, AND SCORED IN EXACTLY THE SAME MANNER AS TABLES 7 THRU 10.

TOW GUNNERY TABLE 12: PLATOON GUNNERY QUALIFICATION

For use of this form, see FM 3-22.34. The proponent agency is TRADOC.

SECTION 1:	TOW CREW 1:	TC: ______	LDR: ______	GNR: ______	DVR: ______	DATE: ______
	TOW CREW 2:	TC: ______	LDR: ______	GNR: ______	DVR: ______	
SECTION 2:	TOW CREW 1:	TC: ______	LDR: ______	GNR: ______	DVR: ______	
	TOW CREW 2:	TC: ______	LDR: ______	GNR: ______	DVR: ______	TIME: ______
SECTION 3:	TOW CREW 1:	TC: ______	LDR: ______	GNR: ______	DVR: ______	
	TOW CREW 2:	TC: ______	LDR: ______	GNR: ______	DVR: ______	

PLATOON: ______ COMPANY: ______ BATTALION: ______

TASK	CONDITIONS/ TARGETS/ SITUATION	TOTAL TARGET EXPOSURE TIME	HITS	ENGAGEMENT TIMES	POINTS	CREW/ LDR CUTS	TOTAL POINTS
1. ENGAGE A SINGLE TARGET	FIRED FROM BASELINE. 3,000 - 3,750 METERS 2 - MV TANKS, 2 - ST BMPS	TFTT BMP/25 SECS TANK/1:13	1 2 3 4				
2. ENGAGE MULTIPLE TARGETS	FIRED FROM PHASE LINE 2. 2,000 - 3,000 METERS 3 - ST BMPS, 3 - MV TANKS	TFTT HMMWV/2:28	1 2 3 4 5 6				
3. ENGAGE MULTIPLE TARGETS	FIRED FROM PHASE LINE 3. 1,000 - 2,000 METERS 3 - ST TANKS, 3 - MV BMPS	TFTT HMMWV/1:08	1 2 3 4 5 6				
4. ENGAGE MULTIPLE TARGETS	FIRED FROM PHASE LINE 2. 1,500 - 2,500 METERS 2- MV TANKS, 2 - MV BMPS (NBC)	TFTT BMP/34 SECS TANK/1:03	1 2 3 4				
5. ENGAGE MULTIPLE TARGETS	FIRED FROM BASELINE. 3,000 - 3,750 METERS 2 - ST TANKS, 2 - ST BMPS (NBC)	TFTT 25 SECS ALL VEHICLES	1 2 3 4				
					TOTAL SCORE		

NOTES:

1. TABLE 12 IS CONDUCTED, TIMED, AND SCORED IN EXACTLY THE SAME MANNER AS TABLES 7 THRU 10.

DA FORM 7341-R, AUG 2003

APD V1.00

Made in the USA
Monee, IL
05 November 2026

67980383R00184